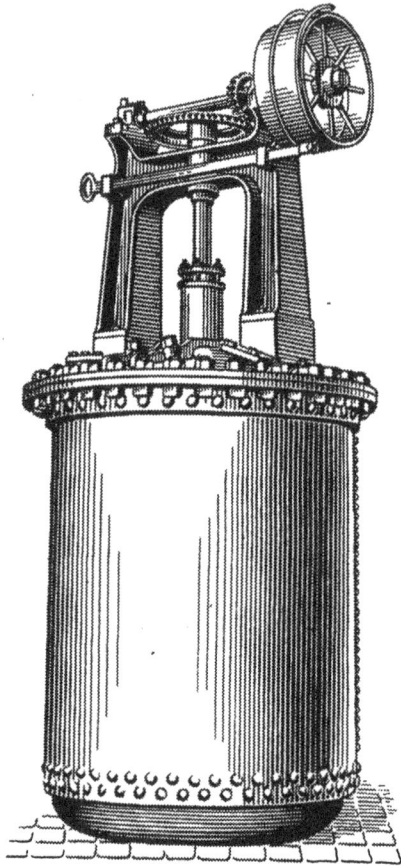

A Typical Low Pressure Works Autoclave.

Laboratory Autoclaves, High Pressure and Hydrogenation Apparatus

Design and Construction

BY

HAROLD GOODWIN, M.Sc.

Wexford College Press

2006

CONTENTS

CHAPTER I

1 *   ix

LIST OF ILLUSTRATIONS

xi

INTRODUCTORY

CHAPTER I

REGARDED in a general way, an autoclave is a piece of apparatus designed in order to render it possible to heat liquids above their boiling point, or to carry out some chemical reaction in which it is essential that the reacting bodies be under a pressure greater than normal atmospheric pressure. The earliest and simplest forms of autoclave were based on the principle of Papin's digester, but at the present time there is in existence a great variety of types in order to meet the varied requirements of modern chemical processes. Not only are autoclaves made from a number of different materials, but the sizes, general design and arrangements for heating are so varied that one might well forgive a chemist, used only to the most common forms, for failing to recognise some of the rarer types of modern apparatus as being autoclaves at all.

In order to render this work of the greatest possible help it is thought best to outline at this stage the scope of the book and the system on which it is written. In the first place it is intended to be of an essentially practical nature, being designed not only to describe the most important types of apparatus or plant used on high pressure work, but to give to any chemist coming fresh to this branch of technical science just that practical information which is, alas! generally only learned in the school of bitter experience.

It is, therefore, because of this outlook that

very little knowledge of autoclave working is assumed, for, after all, the word " obvious " is really very relative. For example : to a man skilled in the firing of pans it is " obvious " that a batch may be maintained at a desired temperature, to within one degree, for hours by the careful adjustment of the fire-box door, but the author has seen many a batch ruined through lack of this elementary piece of technical knowledge.

It is proposed to deal first in a general way with the most common construction of autoclaves, to indicate essential features and to lay down the principles of high pressure work.

Next some space will be devoted to apparatus especially designed for laboratory working both as regards construction and use, before dealing with what is perhaps the most important branch of all—plant used for manufacture whether on a large or small scale. There is, of course, such a variety of designs of pressure pans that it would be quite outside the scope of this book to deal fully with every kind, but two particular types of autoclaves have been selected as covering the most important ranges of high pressure work. These are

(1) The high pressure agitator autoclave, and
(2) The low pressure agitator autoclave of large size.

Separate chapters will be given to each of these pans, in which the plant will be described fully and any points of special interest in their working will be emphasised.

The next section will be devoted to the difficulties met with in the working of autoclaves while carrying out a variety of chemical reactions, and how best these difficulties can be obviated or at any rate lessened.

In conclusion, a chapter will be given to the economical working of numbers of autoclaves so as to utilise to best advantage the capital sunk in such plant, and to those precautions which must be taken when one is in charge of high pressure pans, not only to ensure a minimum of wear-and-tear losses, but, what is far more important, to safeguard in as far as human thought can, the lives and health of the men engaged in working the plant.

OBSERVATIONS ON AUTOCLAVES IN GENERAL

Autoclaves can be made of cast iron, steel, copper, bronze or tin, but some kind of steel is by far the most common material used in their manufacture on account of its great strength. Modern practice supports the use of nickel steel or nickel chrome steel for those autoclaves designed to withstand very high pressures.

Autoclaves are made over a very large range of sizes, from small laboratory pieces of apparatus of a few hundred cubic centimetres' capacity to huge pans capable of holding a charge of two thousand gallons. Certain sizes, however, have been found by practice to be most convenient, and the following table is intended to show the most general limits of different types of autoclaves as

18 *Autoclaves and High Pressure Work*

regards capacity, temperature of heating, and pressure. These figures must not be regarded as rigid limits, but rather as an indication of the extremes fixed by general experience. Special cases always require special treatment, and these tables would not be of much help had they been made to include all possible examples.

TABLE I

Showing the General Limits of Various Types of Autoclaves

Type of Autoclave.	Capacity.	Temp. Limit.	Pressure Limit.
Laboratory non-agitator type.	500 to 1000 c.c.	500° C.	500 atmos.
Laboratory non-agitator type, tubular shape.	300 to 500 c.c.	1500° C.	1000 atmos.
Laboratory agitator type.	1000 to 2000 c.c.	300° C.	300 atmos.
Semi-large-scale autoclave, agitator type (a)	5 gallons	300° C.	200 atm.
(b)	20 gallons	300° C.	50 atm.
Large-scale autoclave, non-agitator type.	200 to 400 gallons	300° C.	100 atmos.
Large-scale autoclave, low pressure agitator type.	700 gallons	200° C.	10 atmos.
Large-scale autoclave, high pressure agitator type.	400 gallons	300° C.	40 atmos.
Special type horizontal semi-large-scale agitator type steam-jacket heating.	500 gallons	100° C.	40 atmos.

The capacity figures refer to the actual capacity of the apparatus, not to the capacity of the charge.
The pressure figures refer to the actual working pressure.

Autoclaves are generally cylindrical in shape, the height being from two to three times the diameter. The bottom is hemispherical in the laboratory types, but rather more like a shallow dish in the larger kinds of pans. In order to give strength, all sharp curves or angles are excluded from the design. Wherever possible the body of the autoclave is cast in one piece, although in the manufacture of works autoclaves it is impossible to avoid joints. This will be dealt with more fully when these pans are being considered in detail, but it is enough to note here that, although there seems to be no theoretical objection to welding, most chemists who have worked autoclaves prefer riveted ones. The top of the autoclave, known as the cover, is fixed on to a flange, cast for that purpose in the body of the pan, by means of nuts and bolts. The joint between the cover and the body of the autoclave is made against a ring, sunk in a groove in the flanged top of the body, the joint being either half or full register.

Many different materials are used for packing this joint, which is often a source of leakage. It is therefore of the greatest importance to have it made as gas-tight as possible. The commonest packings are lead, copper, aluminium and asbestos. Not only does the choice of material used depend on the type of autoclave, but also largely on the pressure it has to withstand and the nature of the chemical operation to be carried out inside it.

In the covers of all autoclaves are certain holes,

generally tapped with screw threads for particular fittings. The number of these fittings varies with the type of autoclave, but there are, however, certain essentially necessary fittings common to all autoclaves—such as the manometer, to register pressure, and the thermometer-tube. The latter usually contains oil in which the thermometer stands, in order to give a more accurate and rapid temperature reading of the contents of the autoclave.

It was at one time the custom to provide two manometers and two thermometers, especially in large autoclaves, the idea being to give some indication of local action if any took place, and as a proof of the efficiency or otherwise of the agitator. As, however, every joint is a possible source of weakness, modern practice is rather in favour of one manometer and one thermometer-tube.

Again, it was common at one time to fix one or more safety-valves of the ordinary steam boiler type to an autoclave, but these were constantly getting choked with material, either distilled or splashed into the valve, which rendered them useless. Devices specially designed for autoclaves are now taking the place of the ordinary safety valves. The most common source of leakage in an agitator autoclave is the stuffing-box, the correct packing of which is a most important point. The construction of the stuffing-box has a considerable influence on its efficiency. It is generally kept cool by a water-jacket, although this is not essential for low pressures—say up to 30 atmospheres. It is

very necessary that the stuffing-box be of ample
size, and still more necessary that it be easily
accessible for repacking. Often a slight leak
develops when the full pressure is reached, which
can be overcome by tightening up the nuts that
control the packing of the stuffing-box.

It should be observed that these are the only
nuts and bolts which it is really safe to tighten
under pressure, for, although familiarity breeds
contempt and workmen often *do* tighten up the
bolts of the cover or manhole lid, it is a dangerous
practice, as the sudden stripping of a screw thread
might lead to disastrous results to the operator.

Before concluding these general remarks, some
space must be given to the question of linings.
Theoretically, the charge should never come in
contact with the inside of the walls of an autoclave, but, in practice, this excellent advice is
seldom taken. Indeed, where there is no possibility of chemical action between the charge and
the material of which the autoclave is made, it is
doubtful if the use of a liner is really a necessity.
Still, in many cases repeated use causes corrosion,
which undoubtedly shortens the life of the autoclave. The liner is a thin-walled vessel made to
fit the inside of the autoclave and intended to hold
the charge.

The liners may be made of lead, sheet tin, copper,
iron (plain or enamelled) or sometimes zinc. The
fixing in position of the liner is a somewhat difficult
operation and one which it is essential to carry out
correctly, for if there be any air gaps between the

liner and the autoclave walls the transference of
heat will be imperfect. This will lead to the walls
of the autoclave becoming too hot in places and may
even result in the splitting of the riveted seams in
large-scale pans. In order to overcome this,
some conducting material is used between the
liner and the autoclave walls, just as oil is used
in the thermometer-tube. Paraffin wax has been
used with some success, but has become almost
entirely superseded by solder. The autoclave is
heated slightly above the melting point of the
solder and the liner placed in position. Solder is
melted and poured round the liner, which has
previously been filled with boiling water. When
a satisfactory packing of solder has been obtained,
the water in the liner is cooled by means of a coil.
The question of the heating of autoclaves will be
considered more fully in subsequent chapters, but
it is interesting to note here that where liners are
used, the oil-bath for heating the autoclave becomes
almost a necessity, otherwise the local superheating
of the autoclave walls causes the solder to melt and
the liner to rise and jam against the inside of the
cover.

With regard to manometers, the one which finds
almost universal use is the steel-tube type. Some
are made with bronze tubes, but these possess no
advantages and are inadmissible in operations in
which ammonia is evolved. In certain reactions
substances are sublimed or steam-distilled into the
manometer and cause a choke in the steel tube
itself or in the pipe leading to the manometer. This

pipe is made wide and often looped to form a trap for such material, but in bad cases there is no cure, and the chemist has to be content with the temperature control. This naturally introduces an element of danger, since one is working in the dark as to the internal pressure, and if there is any possibility of reactions setting in which would cause abnormal pressure some safeguard must be taken. The use of a safety valve of the ordinary type would be of little value, as this would probably choke exactly as does the manometer. The only possible working arrangement would be the fitting of a wide exhaust pipe closed by a cap which would blow off at the safety limit. The operating of the device would probably mean some loss of material, but it would, at any rate, give some feeling of security. Contingencies of this description are fortunately rare.

LABORATORY AUTOCLAVES—
CONSTRUCTION

CHAPTER II

THE laboratory autoclave, as the name implies, is a piece of apparatus designed for the carrying out of reactions under pressure on a small scale. Although, of course, there is very little difference in size between the largest laboratory autoclaves and the smallest semi-large-scale plant, for convenience of classification it is best to regard the laboratory autoclave as designed for experimental work rather than manufacture on even a very small scale.

Manufacturers of laboratory autoclaves put on the market a very large range of apparatus in order to meet all requirements, but, as it would be impossible in a work of this size to describe in detail every kind, there are certain ones which may be taken as characteristic types. These find very general use in all classes of chemical work, and serve to show the fundamental principles of laboratory autoclave construction.

The following pieces of apparatus will therefore be considered as illustrating very representative examples :

(1) Small non-agitator autoclaves.

(2) Small agitator autoclaves.

(3) Large agitator autoclaves.

(4) Medium sized non-agitator autoclaves specially designed for uniform oil-bath heating.

(5) Autoclaves for very high pressures and temperatures.

(6) Autoclaves designed for special purposes.

27

Type No. 1.—The small non-agitator autoclave is probably the most common type of all and the one which finds most general use. It is a cylindrical steel vessel, cast in one piece and fitted with a cover, which is screwed down on to the flange of the body by means of nuts and bolts. Fig. 1 shows the general construction. It will be observed that the bottom is hemispherical in order to increase the strength. The diagram is a sectional one, in which the part (*a*) represents a raised ring of steel on the underside of the cover. This fits into the sunk groove (*b*) in the top of the autoclave, at the bottom of which sunk groove is a ring of lead or copper. Hence the result of the pressure exerted when the large nuts (*c*) are screwed down will be to press the ring (*a*) into the comparatively soft metal of the washer, be it copper or lead. By this means a perfect joint is made capable of withstanding any pressures up to the limits for the particular autoclave. The thermometer pipe (*d*) is screwed through the cover of the autoclave by means of a spanner working on the top of the pipe, which is made in the form of a hexagonal nut (*e*). This is done by the manufacturers and, except in rare cases, need never be interfered with. The pipe (*f*), also screwed through the cover, represents the outlet to the pressure gauge or manometer. It is sometimes simply a straight steel pipe of internal diameter $\frac{1}{2}$ inch to $\frac{3}{4}$ inch, or it may be looped in a complete circle, the latter idea being supposed to render choking less likely.

This type of autoclave is made in a range of

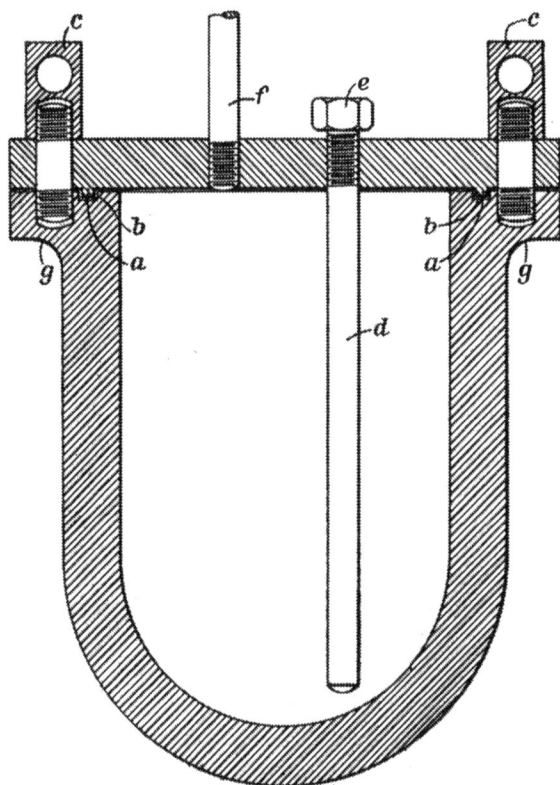

FIG. 1.—Section of a Non-agitator Laboratory Autoclave.

sizes, but the most common is about 600 c.c., which allows a charge of half a litre comfortably. They are suitable for pressures up to 500 atmospheres and temperatures of 500° C., but these limits are well outside the ordinary range of organic chemical work. It must be remembered, too, that special manometers suitable for these high pressures must be employed if such conditions are to be carried out.

Generally speaking, the ordinary limits of chemical reactions carried out in these autoclaves would be up to 300° C. and 60 atmospheres, and a gauge capable of standing 100 atmospheres with perfect safety would accordingly be employed. Most gauges have a red line marking the safety limit, and as the manufacturers allow a very liberal margin of safety, there is no risk if pressures are kept within these limits.

The number of cylindrical nuts (c) employed in an autoclave of the size described is generally six or eight.

This type of autoclave can be heated, either by direct flame, by a heated oil- or solder-bath, or electrically, although the latter method is not usual. When direct flame is employed it is important to use a proper form of iron tripod or metal case, the autoclave resting on a ring which supports the edge of the flange (g). The bottom of the autoclave must be well above the gas ring or whatever type of burner is used, so that it is really heated in an air-bath, which somewhat reduced the chances of burning at the bottom and sides.

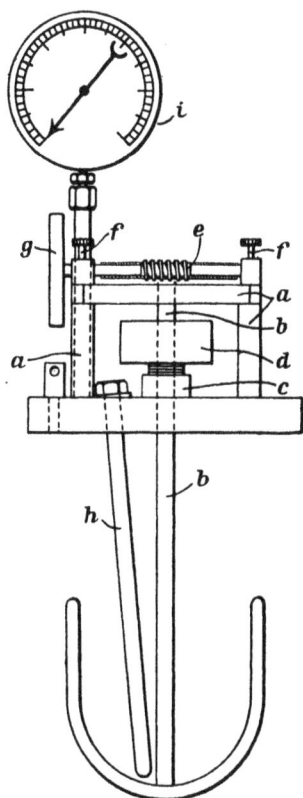

FIG. 2.—Arrangement of Geared Agitator on Cover
of Small Agitator Laboratory Autoclave.

31

If an oil-bath is used it should be a properly constructed cylindrically shaped one, in which the autoclave is supported on its flanged edge, and with some arrangement whereby it can be lifted out of the oil and allowed to drain over the bath.

For oil-bath work, however, the most satisfactory design is that to be described later under type No. 4.

Type No. 2.—This class may be said to embrace laboratory agitator autoclaves up to 1 litre capacity, the most general sizes being 600 c.c. and 1 litre. The general construction of the body is exactly similar to the small non-agitator type, and a satisfactory joint is made between the cover and the body-flange by means of the raised circular ridge working in a groove against a lead or copper washer, exactly as described previously.

Fig. 2 shows the arrangement of the top of the cover.

The general superstructure to bear the agitator shaft is shown by (*a*), which is so placed that the shaft (*b*) passes through the centre of the cover. The shaft terminates at the bottom in the agitator blade, which is often of the anchor type, although many other designs are in use for special requirements. The shaft passes through the top of the cover into a heavy piece of metal (*c*), which is either cast as part of the cover, or screws into it. This piece of metal is to receive the stuffing-box (*d*), which is packed with asbestos, copper, lead or some such suitable material. A liberal size of stuffing-box should be allowed, and it should be possible to tighten up after heating has commenced by screwing it further down. After passing through the stuffing-box, the shaft is keyed on to the cog-wheel, which is operated by the worm-drive (*e*). This driving shaft is lubricated by vaseline cups (*f*) at either end, and power is applied by means of a pulley (*g*) driven by round leather belting from shafting or a small electric motor. The arrangement of the thermometer tube (*h*) and manometer (*i*) requires no explanation.

In many of the first-class models all rotatory parts are fitted with ball bearings, while two other points of great structural importance are, (i), the agitator should almost scrape the sides and bottom of the autoclave, and (ii) the thermometer-tube should be as long as possible.

These autoclaves can be heated by direct flame, in which case a properly constructed sheet-steel surround with furnace-door is necessary, the heat being supplied by a large gas-ring or Fletcher burner. A more uniform heating is, however, obtained if an oil-bath is used. To reduce the possibility of local over-heating to a minimum, some models have an arrangement whereby the oil itself is agitated by means of a plunger driven by a separate pulley off the same shafting which operates the stirrer.

Type No. 3.—The large-size agitator autoclave designed for laboratory work can be obtained in a great variety of sizes, but the most useful is that which will comfortably take a charge of 2 litres. This size not only enables a chemist to make a

2

reasonable quantity of material for subsequent research work, but enables him also to try out a recipe on such a scale as to give him some idea as to how the process will go on the large scale.

The general construction of this type of autoclave is so similar to the smaller one described under Type No 2 that a detailed description is unnecessary, Fig. 3 being self-explanatory. It will be seen that there is no fundamental difference from the smaller models in so far as the arrangements for agitation are concerned. These larger autoclaves are generally fitted with some form of valve for releasing the pressure, the arrangement shown in the diagram being usual and, on the whole, efficient.

It will be seen that a common pipe serves as the outlet to this valve and to the pressure gauge. The advantage of having this valve fixed is two-fold. Not only does it form a means of releasing pressure should anything abnormal occur, but it can be used in certain classes of experiment for blowing off steam at the end of the reaction, thereby accelerating the rate of cooling. Although one might expect these valves to block up, in practice they keep wonderfully clear, nor are they a source of leakage if reasonable attention is given to them.

The diagram shows an autoclave standing in a furnace for " direct-fire " heating, generally by means of gas rings. They can, however, be heated by electricity or in an oil-bath.

The modern practice of good firms is to supply the gas furnaces lined with fire-bricks, with a brick

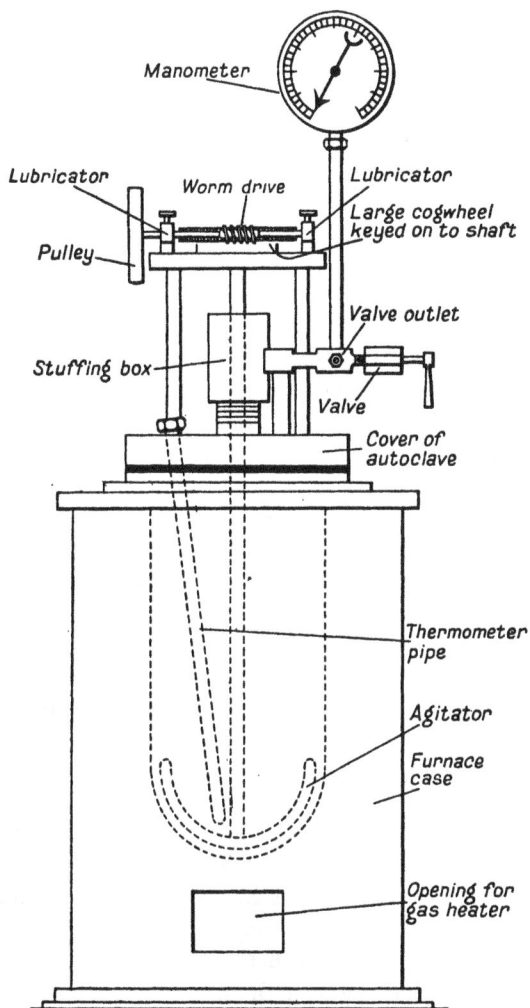

FIG. 3.—Large Agitator Laboratory Autoclave.

35

baffle under the flue fitted to the back of the furnace. The gas burner is such that no direct flames impinge on the bottom or sides of the autoclave. This type of autoclave is really an almost perfect model of a works high-pressure agitator pan, and is capable of standing temperatures up to 300° C., and pressures of 200 to 300 atmospheres.

Type No. 4.—For certain reactions, it may be necessary to keep the whole of the autoclave and cover at a uniform temperature, and under these conditions the reaction will proceed quite satisfactorily without agitation.

It is for such cases that the special oil-bath autoclave about to be described is designed. The capacity of the autoclaves varies from 500 c.c. to 2 or 3 litres, but the 1 litre size is the most useful. In general construction it is similar to the standard non-agitator autoclaves, except that it is not usually built quite so substantially, being designed more for the limit of 200 atmospheres pressure and temperatures up to 300° C. The method of fitting the cover and the arrangement of manometer and thermometer-tube present no modification. The chief point of interest to observe is, that the autoclave rests on a number of lugs (*a*) cast or riveted to the sides of the oil-bath. This latter is of such a size that the autoclave can be completely immersed in oil or other heating medium to a depth of 1 inch above the cover. The plates (*b*) are of steel and are arranged to swivel in order that the autoclave flange can be made to rest on

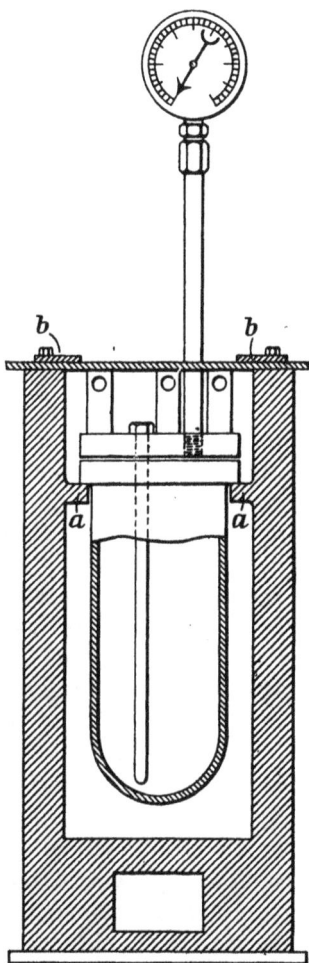

FIG. 4.—Non-Agitator Laboratory Autoclave for
Complete Oil Immersion.

37

them when it is desired to stop the experiment, and withdraw from the bath. This is particularly convenient, as it not only enables most of the surplus oil to drain back into the bath, but also forms a useful method of support when undoing the large round nuts preparatory to removing the cover. The heating of the oil-bath is best performed by means of a large ring burner.

Type No. 5.—It has been shown that the limits of pressure and temperature possible with the types of apparatus previously described are very high, in fact higher than is usually required in research work on synthetic dyestuffs, or general organic chemistry. It is sometimes necessary, however, to carry out reactions under pressures higher even than 500 atmospheres and temperatures above 500° C.

For these cases, special apparatus must be employed which may be regarded as autoclaves, although they are really more in the nature of electric furnaces, especially where very high temperatures are required. No general principles can be laid down for this type of vessel. It must vary greatly according to the kind of work for which it is designed.

One example is a steel vessel of about 3 litres capacity, fitted with adjustable electrodes at the top and bottom. The heating medium is a graphite rod or tube with nickel water-jacketed holders. The whole steel vessel is itself immersed in a water-jacket of cast iron or steel. A valve passes through this water-jacket from the inner steel vessel to the air. Sometimes this valve is fitted with a quartz window so that it is possible to view the interior of the steel vessel when working at high pressures and temperatures.

It is possible to obtain internal temperatures of 1500° C. with this type of autoclave and pressures up to 1000 atmospheres, although it is not safe to go above 500 atmospheres if the quartz window is fitted.

If very high pressures but temperatures not above 500° C. are required, it is possible to use apparatus of more orthodox design in which the heating medium is a gas furnace. The points of interest to note in considering this type of apparatus are (1) that only a very good-quality high-speed tool steel is used in the manufacture of the autoclave, (2) the walls of the autoclave are very thick, (3) the height of the autoclave is very much greater in proportion to its diameter than with ordinary models. This means that for any given capacity the diameter of the ring used to make a joint between the cover and the body is much less, thereby reducing the possibility of leakage. In other words, the apparatus approaches more to the dimensions of a steel tube. Again, all the bolts which screw the cover to the body are of heavier design, and a copper or even a soft steel ring is used to make the joint.

These autoclaves are non-agitator and can be made capable of withstanding pressures up to 1000 atmospheres.

The height of the body is from six to ten times the internal diameter.

Type No. 6.—In addition to those autoclaves specially designed for very high temperatures and pressures, there are numerous models which are normal in so far as their pressure and temperature limits are concerned, but which depart from the standard types in other respects. For example,

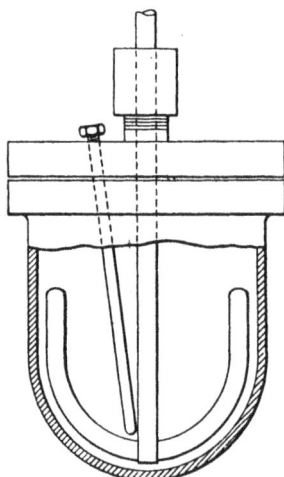

FIG. 5.—Anchor Design   FIG. 6.—Propeller for Stirring
Agitator.   Thin Mixtures.

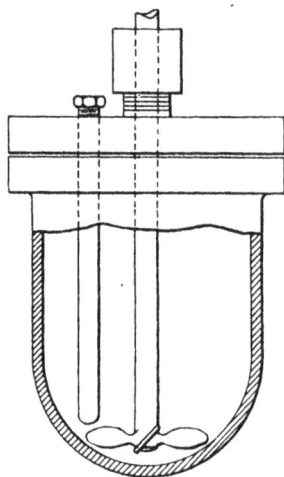

there is the question of agitation. By far the most common type of agitator used in laboratory autoclaves is the anchor design, which gives efficient stirring for all ordinary purposes. A great advantage of this type of agitator is that it can be made to sweep within $\frac{1}{16}$ inch, or less, of the sides, thus preventing the accumulation of solid crusts and consequent burning of the same. Again, the

thermometer pipe can be made to slope within the
space between the shaft and wings of the anchor,
thus getting very near to the bottom of the pan
(Fig. 5). The disadvantage of the anchor type of
agitator is that, with thick pastes, the inside
portions are only imperfectly stirred. For really

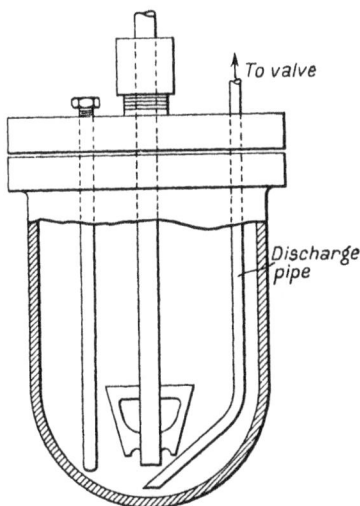

FIG. 7.—Agitator for Reaction requiring
little Stirring.

efficient stirring of thin mixtures, say an oil and an
aqueous layer, there is nothing better than a good
propeller (Fig. 6). This causes up and down
agitation as well as the usual circular whirlpool
kind, but it is not very efficient with thick pastes.
Fig. 7 shows an agitator arrangement for a reaction
in which very little stirring is necessary, but where
2*

it may be required to blow off samples of the batch
from time to time during the course of the experiment.   A further advantage of this type of agitator
is that very small batches can be made, as the
thermometer pipe comes almost to the bottom of
the vessels.

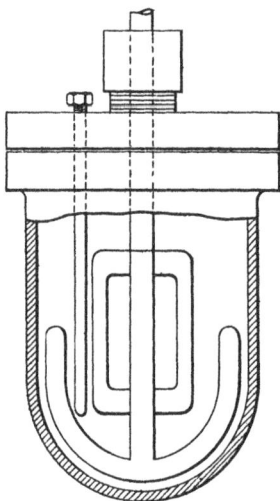

FIG. 8.—Gate Agitator for Stiff Pastes.

For the agitation of stiff pastes a further " gate "
is sometimes fitted to the anchor type (Fig. 8), or
a double agitator autoclave is used.   This latter
is an excellent device for high-speed agitation of
any sort, and is perhaps the best of all systems
where very vigorous agitation is essential and the
pressures are low.   As will be seen in Fig. 9, both
agitators are driven off the same pulley in opposite

directions, the top bevel cog (*x*) operating one, and
the bottom (*y*) the other.

Another type of autoclave worthy of mention is
that with which it is possible to introduce, or to

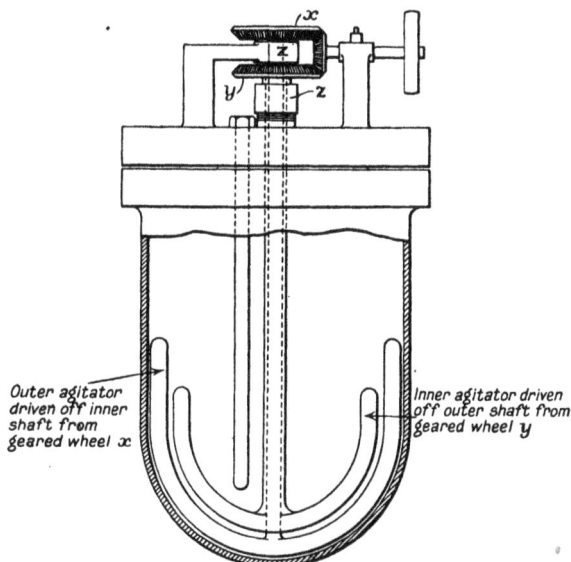

Outer agitator
driven off inner
shaft from
geared wheel *x*

Inner agitator driven
off outer shaft from
geared wheel *y*

Fig. 9.—Double Agitator Type of Stirring.

The glands *z* should be packed so as to stop leakage
not only between the outer shaft and the outside,
but also between the two shafts themselves, otherwise
the contents of the autoclave will be driven up and leak
into the gland. This type of pan is only suitable for
low pressure.

extract, a gas while working at high pressures.
The autoclaves are generally of about 500 c.c.
capacity, of special steel and capable of with-

standing pressures of 500 atmospheres and temperatures up to 500° C.

They are supplied with manometer, thermometer-tube and two valves, connected to inlet and outlet pipes respectively. The inlet pipe passes down to the bottom of the autoclave and the outlet to the top.

There are, of course, many other types of auto-clave, and makers of scientific apparatus are generally quite willing to discuss the requirements of their clients and to design for special needs. For example, the author would mention in this connection the firm of Chas. W. Cook and Son, of Ashby-de-la-Zouch, whose autoclaves he has used with complete satisfaction for many years. Some of the types described in this chapter are illustrative of their standard models.

LABORATORY AUTOCLAVES—USE

CHAPTER III

WHILST there are a great variety of chemical reactions which are best carried out in autoclaves, or for which autoclaves are, if not essential, at any rate desirable, three may be taken as illustrative of high pressure work. It should, however, be explained that the term " high pressure " means pressure above the normal atmospheric pressure, and not necessarily great pressure as measured in lbs. per square inch.

The three chemical reactions to be considered are :

(1) The substitution of a hydroxy-group for a sulphonic acid group.

(2) The substitution of an amino-group for a hydroxy-group or a halogen group.

(3) The introduction of one or more alkyl groups into an amino-group.

The particular types of apparatus most suitable for these reactions differ very greatly one from the other, and it is because of this that these examples have been selected.

(1) The first is the well-known caustic fusion or " melt " process, which is of such great importance in the manufacture of intermediates for azo-dyestuffs. Generally speaking, the process consists in melting a certain calculated quantity of caustic soda or potash, either in the autoclave itself, or in some other suitable vessel for transference to the autoclave. The melted alkali is then raised to a

certain temperature, which naturally varies with the particular process, and the sulphonic acid, say of the naphthalene series, or its sodium salt is charged in. This substance may be in the form of a dried powder or a stiff paste; in any case, it probably contains a large proportion of inorganic salts such as sodium chloride or sulphate. When all has been added, the resultant mixture is a stiff paste, and will probably be a stiff paste even at the temperature of the reaction; moreover, the inorganic salts will be out of solution. These points should be considered in the choice of the autoclave, for to carry out such a reaction in a non-agitator autoclave heated by direct flame would certainly cause local burning and uneven mixing of the charge. Possibly some parts would be melted to dihydroxy-derivatives, if the starting material contained more than one sulphonic acid group, whereas there would be unchanged sulphonic acid in the centre of the autoclave.

By this it is not meant that it would be *impossible* to carry out such operations in non-agitator direct-fired laboratory autoclaves, but rather that such a piece of apparatus would not be desirable, much less ideal.

Since most of the caustic fusion recipes given in such works as Cain's "Manufacture of Intermediate Products for Dyes" for working in autoclaves recommend the use of caustic soda of from 25% to 50% strength, and temperatures below 200° C., it follows that the pressures generated will not be very high. For this type of reaction the pressures rarely exceed 10 atmospheres, so there is absolutely no reason why agitator autoclaves should not be used.

As the paste is fairly stiff, the type of the agitator most likely to do its work thoroughly is the anchor design, supplemented perhaps by an inside " gate " (Figs. 5 and 8). The speed of agitation need not be high—in fact to attempt high speed would be to risk bending the agitator blades, so, all things considered, an anchor agitator auto-clave of capacity, say, $\frac{1}{2}$ to 1 litre, driven by a worm gear, and heated in an oil-bath, would be the most suitable for a piece of research work involving caustic fusions.

Before considering details of closing up, heating, and opening laboratory autoclaves which are much the same whatever the nature of the operation performed in them, let us briefly touch on the special requirements of chemical reactions (2) and (3).

(2) The operation of substituting an amino-group for a hydroxy- or a halogen-group is known as amidation. The usual procedure is to heat the hydroxy- or halogen-derivative with ammonia, either aqueous or alcoholic, with or without sodium bisulphite, at temperatures sufficient to give a pressure of 6 to 10 atmospheres. Typical examples of these reactions can be seen in any organic text-book, the following being fairly representative of this class of work :

(*a*) the preparation of

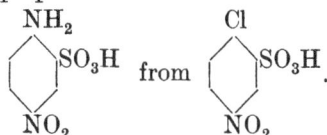

$$\underset{NO_2}{\overset{NH_2}{\bighexagon}} SO_3H \quad \text{from} \quad \underset{NO_2}{\overset{Cl}{\bighexagon}} SO_3H .$$

(*b*) the preparation of

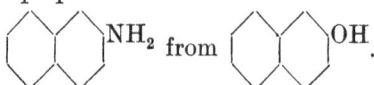

$$\bighexagon\!\bighexagon NH_2 \quad \text{from} \quad \bighexagon\!\bighexagon OH .$$

(*c*) the preparation of

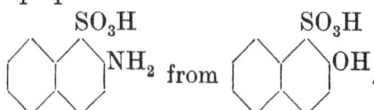

$$\overset{SO_3H}{\bighexagon\!\bighexagon} NH_2 \quad \text{from} \quad \overset{SO_3H}{\bighexagon\!\bighexagon} OH .$$

Here, then, we have a final mixture after charging, which is much thinner than the caustic fusions, and which when on temperature may consist of an oil and an aqueous solution [example (*b*)]. If this operation were carried out in a non-agitator autoclave, poor and inconsistent yields would, without doubt, be obtained, and even if the agitation were of the slow whirlpool type recommended for caustic fusions, it would hardly be sufficient to cope with the needs of the case.

The pressures, again, are not high, although ammonia is one of the most " searching " of all gases, and many a gland which is perfectly tight to steam shows a leak to ammonia.   Still, that can be dealt with, and the pressure will be a diminishing one as the amidation proceeds.   Undoubtedly the type of agitator most suitable for the whipping up of an oil and water is the propeller, although the twin agitator shown in Fig. 9 would probably be satisfactory if the glands did not leak. A worm-drive is not necessary, nor altogether desirable, for bevel cogs would give a greater speed of agitation without having to run the pulley too fast. Since the agitation will keep the contents of the autoclave in constant movement, it would be quite possible to work satisfactorily with direct heating, the advantage being, of course, speed of heating up. There is nothing to be said against oil-bath work, however, if properly carried out, and it is perhaps easier to keep strictly to temperature conditions when an oil-bath is used. On account of the action of ammonia on copper it is essential that this metal be not used, either for packing rings, washers or in the gauge. A lead ring for the cover, asbestos packing for the gland and steel-tube gauges are a satisfactory combination for amidations.

(3) Examples of this type of reactions are as follows :

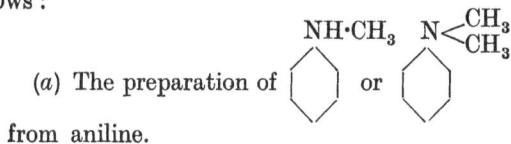

(*a*) The preparation of [NH·CH$_3$ benzene] or [N(CH$_3$)$_2$ benzene] from aniline.

(*b*) The preparation of ethyl α-naphthyl-amine from α-naphthylamine.

Further examples could be quoted in large numbers, as these reactions are simply typical " methylations " or " ethylations."

The most common methods of introducing methyl-or ethyl-groups into an amino-group are treatment of the amino-body either with methyl or ethyl alcohol and a mineral acid, or with the methyl or ethyl esters of mineral acids, such as methyl sulphate or ethyl chloride.

Although many methylations and ethylations can be carried out under ordinary pressure, there are a great many commercial recipes in constant use in the chemical industry which involve the use of autoclaves.

In working recipes involving the use of dilute mineral acid the autoclave must naturally be provided with a liner, and all steel parts protected in order to avoid corrosion of the autoclave and the development of enormous pressures by reason of the hydrogen evolved. If sulphuric acid is used, it is possible to employ lead as a protective material, but in most cases an enamelled cast-iron liner is more satisfactory. The mixture of the amino-compound (or its sulphate or hydrochloride), mineral acid and alcohol often forms a homogeneous whole. Accordingly this type of operation can often be carried out in a non-agitator autoclave just as easily and completely as in an agitator one, which is a great advantage, considering the difficulty in securing adequate protection of the steel autoclave from the searching acid vapours.

Sometimes methylations or ethylations are carried out by heating a mixture of the amine, methyl or ethyl chloride and milk of lime, the function of the latter being to absorb the acid formed in the reaction. In this case, an agitator autoclave would be used, but it would not· be necessary to employ the enamel liner.

In order to follow exactly the procedure most likely to ensure the successful working of laboratory autoclaves, let us take a typical case—say of the caustic fusion of a sulphonic acid· or aminosulphonic acid of the naphthalene series to the corresponding hydroxy-compound—and go over carefully every step taken.

The autoclave chosen for the work will be an agitator one of from 600 to 1000 c.c. capacity, the type of agitator being a simple anchor, or an anchor reinforced with an inside gate. The autoclave will be worm-driven and heated in an oil-bath.

The size of the batch will be regulated so that the complete charge before sealing up occupies three-quarters to four-fifths the total capacity of the autoclaves. This is very important as, owing to the expansion of liquids on heating, if the charge be too great, enormous pressures will be generated when the high temperature is reached.

The author has known of cases in which the screw threads of the manometer and even the cover bolts have been stripped owing to the neglect of this precaution. In carrying out a series of experiments it is important to keep the sizes of the batches approximately constant, for the golden rule in research work is to alter *one* factor, and *one factor only*, at a time. In spite of much that has been written to the contrary, it is quite possible to carry out caustic fusions in a steel autoclave without the use of a liner, particularly if the fusion of the rock caustic with the necessary quantity of water is done in a cast-iron vessel, and, when melted, transferred to the autoclave. If the liner is properly fitted with solder into the body of the autoclave there can, of course, be no possible objection to the use of it.

Let us suppose, then, that the calculated quantity of caustic soda and water has been fused together and transferred to the autoclave, and the sulphonic acid or its sodium salt is weighed out ready for charging. As the resultant mixture will probably be fairly stiff, it will perhaps be advisable to have the autoclave in its oil-bath and some heat applied in order to end up with a mixture at 70° C. or higher. Naturally no hard-and-fast rule can be laid down on this point, for so much depends on the particular recipe, but the final temperature before sealing up should be as low as is compatible with proper mixing of the charge.

When all the sulphonic acid has been added, the mixture will be a thick cream, and the operation of closing the autoclave is now begun. This is a very important stage and one which may easily be carried out incorrectly by inexperienced hands. In the first place, it is important to see that the sunk ring in the body flange of the autoclaves, on to which the raised rim on the cover presses, is in perfect condition and quite clean. This ring is generally made of lead for most laboratory purposes, although for very high pressures copper is better. The cover is placed over the round bolts on to the body flange and shaken about to ensure that it is " sitting " nicely. Then the large round nuts are threaded down until they just begin to grip. Up to this stage nothing but the fingers has been necessary, but now two stout " tommy bars " should be used. One of these is slipped alongside the round nuts and the other threaded through the circular hole in one nut. Pressure is now exerted so as to tighten up that one nut by half a revolution. The bars are withdrawn and the nut immediately opposite to the one first tightened is next turned. In this way all the nuts are in turn screwed down, always going from any nut to the one immediately opposite to it. When all the nuts have been so treated twice, the procedure is changed, and the nuts are screwed down one after the other, passing round and round the cover until it is impossible to screw any nut down any further. It is absolutely unnecessary to use a hammer to tighten up a laboratory autoclave, and such a procedure only risks stripping the thread of the screws. The autoclave is now heated up, the agitator being run at a steady continuous speed. Unless there is some special reason to the contrary, it is as well to heat up to the correct temperature as quickly as possible, which may take any time from one to six hours.

The raising of an autoclave to temperature and the subsequent maintaining of it at that temperature is by no means an easy operation, but one that requires considerable experience. Still, it is possible to give some hints as to how to set about the task without claiming that the method described is the only practicable one.

In the present case let us suppose that the desired internal temperature of the batch is 180° C., which is a fairly representative figure for this class of chemical action. The temperature of the oil-bath containing the autoclave would be raised steadily to this figure, which might take from one to two hours. As there is always a " lag " or difference in temperature between the oil-bath and the inside of the autoclave, even when both are perfectly constant and steady, there is no risk run in raising the oil-bath straight away to the desired temperature. At this point the inside temperature of the batch will be perhaps 120° C., as shown by the thermometer placed in the thermometer-tube which has previously been partially filled with oil. An attempt is now made to steady the temperature of the oil-bath by adjustment of the gas or electric heating appliances, the object being to slow off the rate of heating of the oil-bath and to bring its temperature to about 10° to 20° C. above the desired internal temperature of the charge. While this is being done, the temperature of the charge in the autoclave rises steadily, and finally comes to a figure about 10° to 40° C. below that of the oil-bath at which it is constant. The difference between the two temperatures when both are steady is the " lag," and depends on the capacity and the design of the apparatus. Where the loss of heat by radiation from oil-bath and autoclave top is low, the " lag " is small, and *vice versa*. Let us suppose we have reached a position of steadiness and equilibrium, the oil-bath being 195° C. and the thermometer in the autoclave tube showing 175° C., *i.e.*, 5° too low. We now slightly increase the heat on the bath and wait till it comes up and remains steady at 200° C. When this has occurred it is probable that there has been no change in the autoclave thermometer reading, but we must be patient, for the heat takes some time to pass from the bath and raise the temperature of the large mass of autoclave and charge. Only when we are sure that the full effect of our heating of the oil-bath has been reflected in the thermometer reading of the autoclave should we consider a further increase or decrease in heat on the bath if required.

It is essential that a chemist should be most systematic and patient if he is to acquire any skill in regulating autoclaves, particularly those heated in oil-baths.

Once the autoclave has been got " on temperature " the maintenance of it at that degree of heat is a simple matter provided one remembers that one must never make sudden and drastic alterations in the heating and must always wait for the effect of any change to show on the autoclave thermometer before making further adjustments. It is surprising how little an alteration in gas supply to the burner is needed in order to cause a rise or fall of 5° on the autoclave thermometer.

It is advisable to keep a record of the progress of the experiment in the form of a chart, the following being a typical example.

Time.	Temp. of oil-bath.	Temp. of charge.	Pressure.	Remarks.

As soon as the time for the heating of the charge at full temperature is over the heat is turned off and the autoclave is allowed to cool. The practice of cooling very quickly in water is not to be recommended except in cases of emergency, as it is liable to cause strains in the steel of the autoclave. In addition, it is a procedure which it would be impossible to imitate on a large scale. It is, however, quite reasonable to lift the autoclave from its bath and allow it to cool in the air of the laboratory. The chart should be filled up during cooling, and this may yield valuable evidence in some experiments.

As regards opening the autoclave, this is an operation of some danger unless done correctly, especially if through the evolution of a gas during the reaction there is a residual pressure shown on the gauge even when the autoclave is quite cold. Nevertheless if carried out properly accidents need never occur. The first step is to insert the " tommy bars " which were used for tightening up, one between the big round nuts to steady the autoclave on its stand, and the other through one particular nut. Reference to the diagrams will show these nuts to have circular holes in them and the " tommy bars " are made just to fit these holes tightly. Although it was possible to tighten up an autoclave cover without the use of a hammer, it is generally necessary to hit the tommy bar smartly with a 2 lb. hammer in order to loosen the nut. The moment the nut is loosened *ever so slightly* it is left and the operation carried out on the adjacent nut. In this manner all the nuts are loosened, but unless there is a huge residual pressure—say 20 atmospheres—the cover will be so embedded in the lead ring that there will be no apparent diminution in the pressure on the gauge. When all the nuts have been loosened they are unscrewed slightly—say half a revolution—and the edge of a well-sharpened cold chisel is placed between the flange of the cover and that of the autoclave. On driving this home with a few smart blows the cover is eased in the lead ring and the gas causing the residual pressure, if any, escapes harmlessly without splashing the charge over the operator. Even if there is no residual pressure shown, this procedure should be *always* adopted, as often there is a slight puff of gas when the cover is eased which would mean the probable loss of some of the charge if the cover should be taken off carelessly. Once the operator has assured himself that the cover is eased and that any pressure in the autoclave has been relieved, then the large nuts are quickly unscrewed and the cover is entirely lifted off.

The author has opened autoclaves showing from 300 to 400 lb. per sq. inch residual pressure with perfect safety by adopting the method just described.

To sum up, it should be observed that there are certain golden rules governing the successful use of laboratory autoclaves, and although skilled and experienced chemists may depart from these and still obtain satisfactory results, that does not alter the fact that by so doing they are taking a risk which may lead to unreliable and untrustworthy work.

These rules can be stated briefly as follows :

(1) Employ an agitator autoclave wherever possible, and *always* where the starting mixture or final reaction product is not homogeneous.

(2) Employ an oil-bath as the heating medium wherever possible, and always where a solid is likely to separate out during the course of the reaction.

(3) Keep accurate records of the temperature of the oil-bath, temperature of the mixture and pressure. These observations should be charted every half-hour.

(4) Try to regulate the size of the batches to occupy 70% to 80% of the capacity of the autoclave when fully charged before heating up.

(5) Remember that time spent in correctly tightening up the bolts which hold the cover to the body of the autoclave so as to ensure an even pressure on the ring is well spent.

(6) Get the mixture to the proper reaction temperature as quickly as possible without taking risk of burning. It is safer to force an agitator autoclave than a non-agitator. In special cases it may be necessary to heat up slowly in order to cause a reaction to take place over a range of temperature to avoid getting too high pressures.

(7) Never tighten up any nuts and bolts except those on the stuffing-boxes when the contents of the autoclave are under pressure.

(8) Never depart from the correct method of opening an autoclave as previously elaborated. It is perfectly safe to open an autoclave with residual pressure inside if the proper procedure is adopted, but the careless opening of one containing caustic soda may cause the operator to lose his sight.

SEMI-LARGE-SCALE PLANT

CHAPTER IV

THERE is some difficulty in correctly defining semi-large-scale plant, as there is bound to be a gradual transition from the very large laboratory autoclave to the real large-scale plant. It is hardly enough to fix limits in definite capacities, nor is it altogether correct to describe " laboratory " autoclaves as those used for experimental work and " plant " as those used for production.

There are some fine chemicals the market requirements of which are so small that they can be, and are, actually manufactured by the trade in laboratory apparatus, while on the other hand it is often the custom to put batches through in semi-large-scale plant simply as experiments in order to determine without great cost how a new process will go under works conditions. Perhaps the best classification and the one which shows the real distinction between laboratory apparatus and small-scale plant is to regard laboratory apparatus as that which can be moved about from place to place, which is capable of being emptied of its charge by hand and whose capacity does not greatly exceed 4 or 5 gallons.

Semi-large-scale plant is, then, apparatus installed permanently in brickwork in a shed which is worked by workmen under the direction of a chemist, and whose function is either to test out new processes on a small scale or to manufacture small batches of material required in only moderate quantities.

3   65

The capacity of semi-large-scale autoclaves is from 5 to 50 gallons, whereas a small-size works pan will be about 200 gallons.

As was found to be the case with laboratory apparatus, so there are a great variety of semi-large-scale pans, made to meet all sorts of requirements. There are four examples, however, which will be considered in detail as being interesting and representative types :

(1) A moderately high pressure non-agitator autoclave.

(2) A moderately high pressure agitator autoclave.

(3) A very high pressure agitator autoclave.

(4) Steam-jacket autoclaves.

Type No. 1.—The non-agitator autoclave is becoming used less and less of late years owing to the great improvement in the design of agitator pans, particularly with regard to the construction of the gland of the agitator shaft. Still, for some purposes, a gas-heated non-agitator pan is all that is required. The pan about to be described is classified as a moderately high pressure one, and in this connection a few words of explanation are necessary. We have seen that with laboratory apparatus it is quite common to have autoclaves, agitator or otherwise, which are capable of standing pressures up to 200 atmospheres. These high pressures are seldom required in chemical research work, still the construction of the apparatus is such that they could easily stand them. Naturally,

when one is dealing with large pans, the possible
sources of leakage are much greater, and hence it
becomes more difficult to get really full-size works
autoclaves to stand pressures which are considered
low in laboratory apparatus. Semi-large-scale
plant is intermediate in this respect.

A glance at the table given on p. 18 illustrates
this point, and we may summarise the matter as
follows :

Ordinary laboratory autoclaves agitator type
will stand pressures up to 300 atmospheres, while
the non-agitator ones can go to 500 atmospheres.

Really " high pressure " laboratory apparatus
can be made to stand 500 to 1000 atmospheres
pressure.

Semi-large-scale plant for *moderately high pressures* will stand working pressures in the region of
50 atmospheres whether agitator or non-agitator
type.

Very high pressure semi-large-scale plant which
is usually of the agitator type will work up to 200
atmospheres.

In the large-scale works class of pan of capacities
200 gallons upwards, non-agitator pans will work
up to 100 atmospheres, although this type of plant
has many disadvantages.

The high pressure works pan of 400 gallons
capacity or more, fitted with stuffing-box agitator,
has a working pressure of 40 to 50 atmospheres,
while the larger size low pressure agitator autoclave
works best up to 10 or 15 atmospheres.

To return to our Type No. 1 non-agitator

autoclave. This is described as a moderately high pressure semi-large-scale pan, and is therefore usually of capacities of 25 to 50 gallons and working pressures of 40 atmospheres.

Fig. 10 gives a general idea of the type and suitable setting of these autoclaves. As the construction of the pan is so similar to the agitator ones which will be shown later in section drawing, it was thought that a general view of the arrangement of this pan would be more interesting.

The autoclave is machined from solid mild steel forgings and fitted with pressure gauge, safety valve, and thermometer pipe. It is set in brickwork and heated by gas burners, which are the most satisfactory and usual means of heating semi-large-scale plant. In addition to the regulation of the gas pressure on the burners, some regulation of temperature is effected by means of the damper controlled by chain and wheel as illustrated in the diagram.

Type No. 2.—The moderately high pressure agitator autoclave about to be described is a very useful and efficient piece of plant, and represents the latest modern practice in autoclave construction.

Fig. 11 shows a section of this pan and serves to illustrate the outstanding features of the same.

The autoclave is made of cast steel and is immersed in an oil-bath for uniform heating. As with most semi-large-scale plant, gas firing takes the place of the ordinary furnace. The capacity of this type of pan ranges from 20 to 50 gallons and the best working pressures are up to 40 atmospheres,

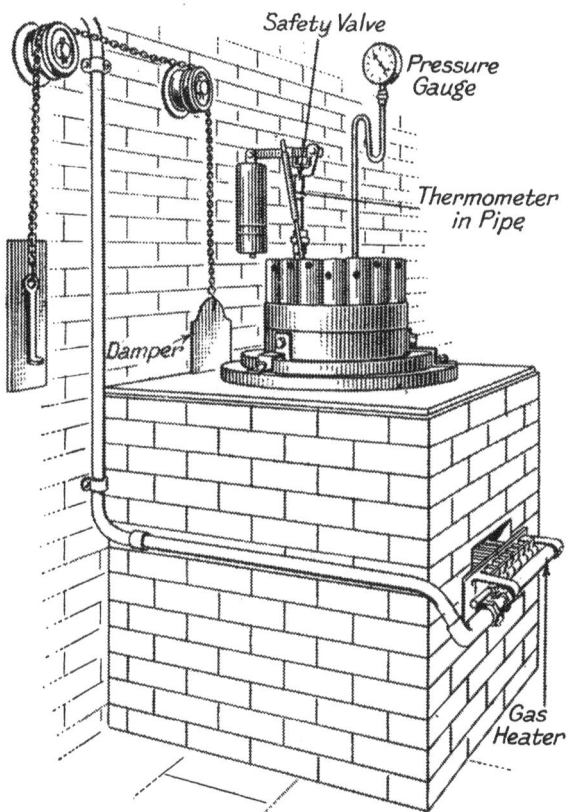

FIG. 10.—General arrangement and setting of moderately High Pressure Non-agitator Autoclave.

although it would be quite safe to work up to 50 atmospheres.

The cover is fitted to the flange of the body by a copper gasket in a full register joint, contact being made good by means of hexagonal nuts and bolts. The agitator is of the anchor type and passes through a water-cooled stuffing-box packed with black-leaded asbestos or lead rings.

Power is applied to the pan by means of a flat belt passing over a " fast and loose " pulley, which renders it possible to stop the agitation at a moment's notice by sliding the belt on to the loose pulley. A good speed agitation is secured by means of bevel cogs.

Other points of interest are the use of a liner to protect the autoclave body, the space between the liner and autoclave walls being filled with solder, and the fitting of the safety collar. It has been found that when working at pressures over 20 atmospheres or even at lower pressures in large-size autoclaves, there is a tendency for the agitator to be forced upwards. This not only causes grinding and wear on the bevel cogs, but also tends to cause leakage through the stuffing-box. The safety collar fixed to the shaft just below the inside of the cover takes this strain and so ensures smooth running. Although no safety valve is shown in the diagram, it is usual to fit one, but the question as to the most suitable kind will be dealt with under the section devoted to large-scale plant, as this is a problem of much greater concern to the users of large-scale plant.

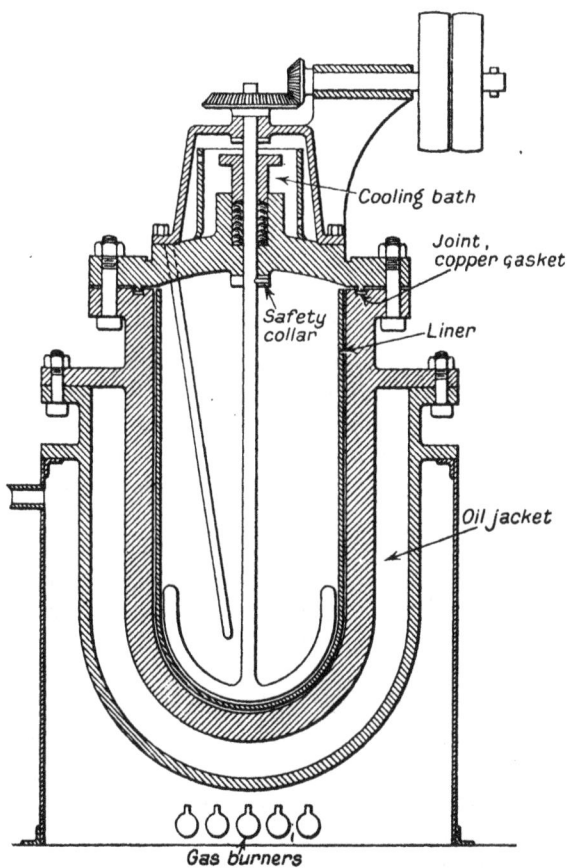

Cooling bath

Joint,
copper gasket

Safety
collar

Liner

Oil jacket

Gas burners

FIG. 11.-- High Pressure Gas Fire Autoclave Section.

71

Type No. 3.—There are now on the market a number of small semi-large-scale autoclaves capable of standing really high pressures and of the agitator type. The pressures referred to are up to 200 atmospheres and the most convenient capacity 5 gallons, although it is possible to obtain larger sizes. Naturally when enormous pressures of this description have to be resisted, all parts of the plant must be somewhat massive, and although the section in Fig. 12 is not intended to be strictly to scale, it serves to show how liberal must be the allowance of metal in autoclaves of this type. To anyone familiar with autoclaves, the diagram itself will show most of the points of interest in this model, but a very brief survey will not, perhaps, be out of place.

The body of the autoclave (*a*), which is cast in one piece, rests on an iron platform (*y*) supported on brickwork or a specially constructed furnace casing indicated by (*x*). Such small pans are usually directly heated by gas, precautions being taken to avoid the play of the burner flames on the bottom or sides of the autoclave. Oil-bath heating as shown for the lower pressure type in Fig. 11 would, however, be perfectly satisfactory and in many ways superior to the most perfect system of direct-firing with baffle plates. The cover of the autoclave (*b*) is of interest, as the shape of the inside is such as to avoid sharp angles, a very necessary thing for semi-large-scale plant, which has to withstand very high pressures. The cover is fastened down on to the body by means of the

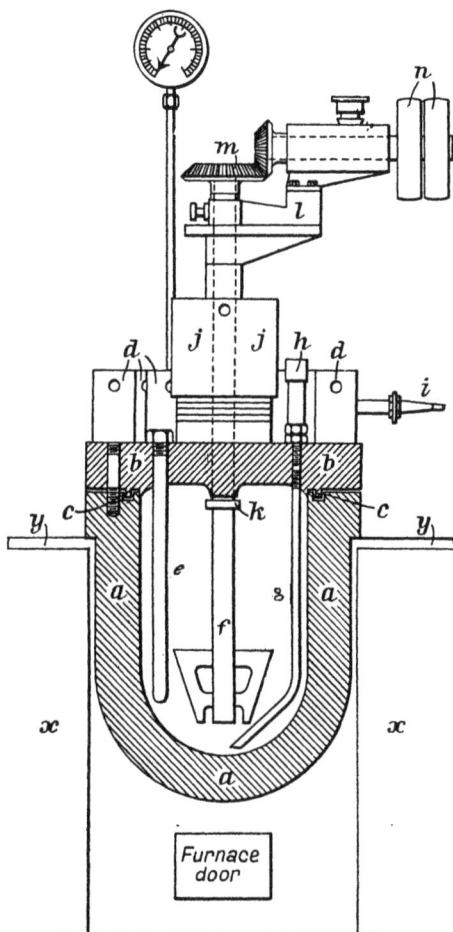

Fig. 12.—Section of High Pressure Agitator
Autoclave—Semi-large-scale Plant.

heavy round nuts (*d*), a perfect joint being made with a full register joint and copper gasket (*c*).

The inside fittings of the autoclave are as usual, viz. a thermometer pipe (*e*) and agitator (*f*), the shape of the latter being peculiar so as to enable a discharge pipe (*g*) to come to the bottom of the pan. The discharge pipe leads to a heavy valve (*h*) and finally to the outlet (*i*), it being sometimes the custom to have a further valve fitted which acts as an inlet for charging or as a blow-off at the end of the reaction.

The agitator shaft is fitted with a safety collar (*k*) as described previously, this being all the more necessary to take strain off the bevel cogs at such very high pressures.

The stuffing-box (*j*) is not usually water-cooled in the 5 gallon sizes, but for anything much larger it would be better so. The usual arrangements for driving the agitator are shown (*j*, *l*, *m*,), the addition of the fast and loose pulleys (*n*) being a very necessary safety precaution in semi-large-scale plant.

Although small in size and somewhat inefficient in agitation, this type of autoclave has numerous advantages and will be found capable of very faithful service for experimental work and small-scale manufacture where high pressures are unavoidable.

Type No. 4. Steam-jacket Autoclaves.—Although there are a great variety of special designs of semi-large-scale autoclaves, too numerous to mention in detail, no section on this subject would be complete without a few words on steam-jacket pans.

As the name implies, these are autoclaves in

which steam is used as the heating medium, sometimes direct from the boiler at high or low pressures
and sometimes superheated where higher temperatures are required.

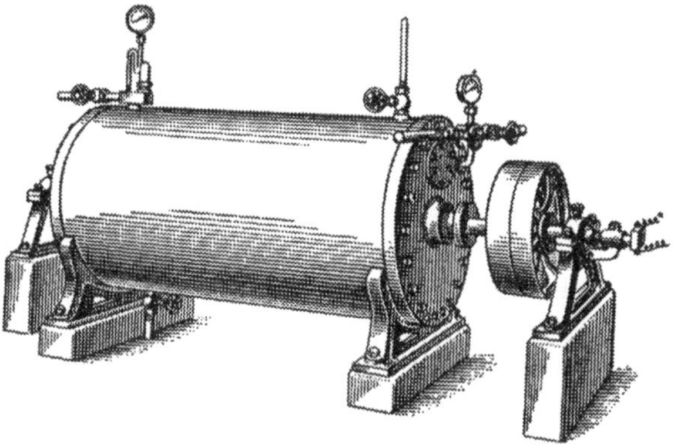

Fig. 13.—Horizontal Steam-jacket Autoclave.

A brief description of two types of steam-heated
autoclave will now follow :

(*a*) Horizontal type—large capacity.

(*b*) Vertical type—small capacity for higher
pressures.

(*a*) A typical horizontal steam-jacket autoclave
is shown in Fig. 13. Although made in smaller
sizes, the most useful is the 50-gallon-capacity pan,
which has a working pressure of 30 to 40 atmospheres. The autoclave itself is made from solid
steel forgings and is fitted with manhole and lid,
pressure gauge, inlet and discharge pipes, together
with the necessary valves. The steam-jacket

takes the form of a mild steel tube fitted into the
autoclave covers at either end.   It is provided with
its own gauge and safety valve, the latter being
of the ordinary steam-boiler type, and inlet and
exhaust pipes for the steam supply.   In order to
prevent loss of heat by radiation it is usual to have
the steam-jacket well lagged with a thick covering of
suitable non-conducting material, such as some of
the asbestos composition pastes sold for steam pipes.

The agitator shaft runs horizontally through the
autoclave from one cover to the other and is fitted
with blades at intervals along its entire length.
It passes through stuffing-boxes at either end, and
is supported also at each end by external bearings.
Power is supplied by means of a flat belt operating
fast and loose pulleys.

The measurement of temperatures in this type
of autoclave presents some difficulty, as a single
thermometer pipe would hardly give a truly representative reading.   The best modern practice is to
use pyrometers which can be arranged to slide
inside the hollow agitator shaft from either end,
thus rendering it possible to obtain accurate
temperature measurement at any point.

(*b*) The vertical steam-jacket autoclave shown
in Fig. 14 is a rather more heavily built type of pan
for pressures up to 60 or 70 atmospheres.   The
general construction of the autoclave is so similar
to designs previously described that no explanation
of the diagram in that respect is necessary.   The
capacity of this autoclave generally ranges from
5 to 20 gallons.   The points of interest are those

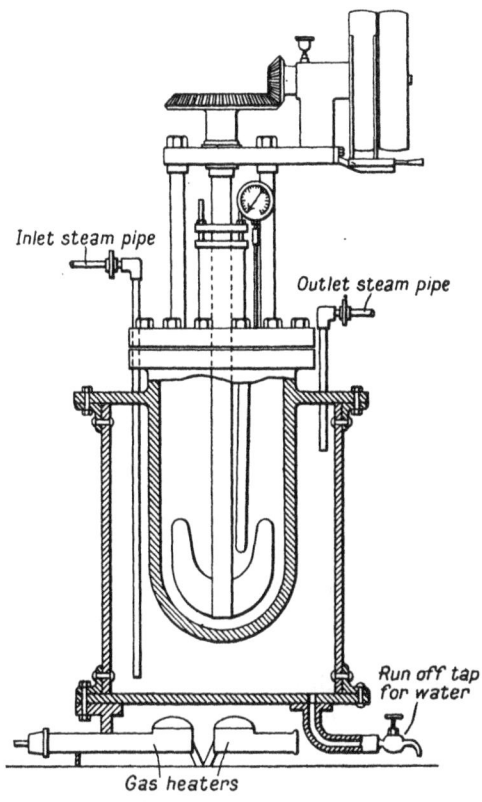

Inlet steam pipe

Outlet steam pipe

Run off tap
for water

Gas heaters

FIG. 14.—Vertical Steam-jacket Autoclave—Section.

connected with the steam-jacket. This is of mild steel, and is fitted with safety valve, steam gauge and the necessary pipes for connecting to the steam mains. In addition, the steam-jacket may be heated by gas burners in order to superheat the steam, by which means temperatures up to 300° C. can be conveniently reached.

If so desired, the inlet valve could be closed and the steam-jacket used as a boiler to generate its own steam.

In concluding this chapter it is not intended to devote any space to the details of working semi-large-scale plant. Although there are without doubt several points of difference, generally speaking the conditions approximate to those of large-scale manufacture, which will be considered at some length in subsequent chapters.

In those models which have *not* permanent blow-over pipes fitted such as in Fig. 12, it is usual to empty the pan by means of a special pipe, which is inserted through the charging hole or through a hole in the manhole lid after the batch is cooled down and the pressure relieved. On passing compressed air into the pan the contents are blown over into any suitable vessel for the next stage of manufacture or experiment. Although it has been impossible to show the compressed air inlet in the diagrams, most semi-large-scale autoclaves except the very smallest sizes have such a fitting, as lifting the cover and bailing out the contents by means of buckets would be a very clumsy and tedious, operation to perform for each batch.

THE CONSTRUCTION AND USE
OF WORKS AUTOCLAVES

CHAPTER V

Introduction.—Up to now we have been considering apparatus essentially designed either simply for experimental work or for very intermittent manufacture possibly of an experimental nature. When dealing with the subject of works plant, we must remember that this is intended for regular routine production and. that therefore several factors which up to now have been of small importance become vital. Foremost amongst these factors is reliability, by which one means that a satisfactory design of works autoclave is one that is capable of withstanding reasonable wear and tear for long periods without constantly needing attention at the hands of the maintenance engineer. When considering the different types of works autoclaves and comparing them with those of the semi-large-scale or laboratory sections, it may seem that some repetition has occurred, for all autoclaves must necessarily be fundamentally similar, but the conditions of works practice are so different that it is desirable to treat this branch of the subject as a separate chapter regardless of what has gone before.

Although works autoclaves may be said to range in size from 25 gallons to 2500 gallons, the lower limits are better regarded as semi-large-scale plant, while the higher are so exceptional as to be outside the scope of this book. For general purposes, works autoclaves may be taken as ranging in size from 200 gallons to 1000 gallons, the larger sizes being those for low pressures, and *vice versa.*

It was stated in the previous chapter that owing to the improvement in the manufacture of agitator autoclaves, the use of the non-agitator type was becoming less common, and this is true of works plant also. For this reason it is not intended to take any particular design of the works non-agitator autoclave into consideration, but as in many old-established works these pans are still doing good service a few general observations on their use will suffice.

The non-agitator autoclave is often a small-sized pan made of very thick steel, and is used for what must be regarded as very high pressure from a works point of view. In some cases the walls of the pan are 4 or 5 inches thick and when properly sealed will stand pressures of 100 atmospheres; indeed the whole problem of getting works pans to stand pressures resolves itself into the problem of getting packed joints to remain good. The non-agitator autoclaves have small manholes for charging—far too small to allow a man to pass through—and the joint of the cover on to the body, once made, is never broken until it is necessary to have the pan cleaned out and examined. This joint is packed either with a lead ring or thick black-leaded asbestos rope, and although the full or half register gasket as in Fig. 11 (Chapter IV) might be employed with advantage, in old pans it is not usual. The nuts and bolts employed to exert the necessary pressure between cover and body are very strong and as numerous as possible. Really satisfactory joints have been made by using rings 1 inch to 2 inches wide and ¼ inch to ½ inch thick, the material being lead, copper, or black-leaded asbestos, and it is good to bear in mind that the method of tightening up and fitting the ring is just as important as the material used. The joint between the manhole lid and the manhole in the pan cover is another source of weakness, but here again a flat black-leaded asbestos ring gives a good joint, although it is advisable to use a fresh ring for each batch.

Non-agitator autoclaves should always be oil-bath-heated unless the contents are an absolutely homogeneous mixture of mobile liquids or a thin aqueous solution. The author once had occasion to investigate variable yields of a caustic fusion of a naphthalene sulphonic acid and found that the operation was being carried out in a direct-fired non-agitator autoclave. No wonder the yields were variable! When working with these autoclaves, it is very necessary to have adequate arrangements for relieving pressures should they get too high. It has been stated that the ordinary safety valve is not very reliable for autoclave work, but, as will be shown later, there are devices which seldom fail. In the event of internal decompositions, if the cover and manhole joints had been made correctly, it is probable that the contents would attain an enormous pressure before anything happened. With no safety device and no friendly agitator gland to protest by leaking and the pressure gauge blocked up, the first intimation of anything wrong would be the " blowing " of the ring off the manhole lid and subsequent ejection of the contents of the pan all over the shed.

Apart from the fact that non-agitator pans are more likely to collect a deposit on the inside and therefore should be cleaned out more frequently, the routine of periodic testing and examination is the same as with agitator pans. It is not always necessary to remove the cover and put men inside the pans for removal of deposits, although this is the last resort. If, however, the works chemist will cultivate the habit of boiling out his pans with water whenever an opportunity presents itself, he will be surprised to find how much dirt and insoluble matter can be dislodged. The cost in firing is very little, and it may mean that the autoclave will continue its routine work for half as long again before having to be dismantled for a complete scraping and examination.

THE LARGE-SIZE LOW PRESSURE WORKS AUTOCLAVE

It is intended to devote some considerable space to the description of the construction of the low pressure works autoclave, as for all-round work there is no doubt that this type of pan is the most useful the works chemist has at his disposal. First of all, without confining ourselves to specific limits, let us decide a reasonable range of capacities and pressures within which to fix the definition of this autoclave. On the works, working with large pans, low pressures may be said to range from 5 to 15 atmospheres, and although with careful attention to the packing of joints it might be possible to work up to 20 atmospheres in the type of autoclave to be considered, it is getting far too near to the actual tested limit to be safe. For regular routine work, by far the best results are obtained at pressures of from 5 to 10 atmospheres, and although these limits may appear close together and on the low side, it is surprising how great a variety of reactions can be carried out at these pressures if the research chemist sets himself to try to devise recipes accordingly. For example, most of the important hydroxy-derivatives of the naphthalene series, particularly aminonaphtholsulphonic acids, can be prepared by recipes which do not involve greater pressures than 100 lb. per square inch, while amidations both in the benzene and naphthalene series seldom give rise to pressures higher than 150 lb. to the square inch, and even that pressure diminishes steadily owing to absorption of ammonia as the reaction proceeds.

It can be taken then, that, although for experimental work we need autoclaves capable of standing many hundreds of atmospheres, the bulk of the most important products required in mass by chemical industry can be made successfully in comparatively low pressure pans.

As regards capacity and temperature limits, there is less need for explanation, as undoubtedly the most useful size of large works autoclave is from

600 to 1000 gallons capacity, while it is seldom that temperatures higher than 250° C. are required.

The frontispiece of this book is from a photograph of a large size low pressure works autoclave manufactured by Messrs. Adamson and Co., Ltd., of Hyde, Manchester. The author has used many of these pans and subjected them to a practical test of reliability by reason of months of unintermittent daily working quite out of the ordinary. They came through the test most creditably and can be confidently quoted as examples of first-class British workmanship. In the following descriptions of a low pressure works autoclave, it is not intended to follow in every detail the Adamson design, but the description and diagram may be taken as substantially correct and characteristic of a works autoclave of this type.

Fig. 15 gives a sectional elevation of the pan, and is worthy of careful study, as it shows marked differences in design from the laboratory and semilarge-scale plant previously considered.

In the first place, the body of the pan, instead of being cast in one piece, consists of a riveted cylindrical tube, made from sheet steel, open at the top to receive the cover, and a shallow dish for the bottom which is riveted on to the cylindrical sides. It is claimed by some authorities that the prejudice in favour of a riveted joint between the dish and the sides is without foundation, but most works chemists who have had experience of both riveted and welded joints will favour the former.

The depth of the cylindrical part of the autoclave

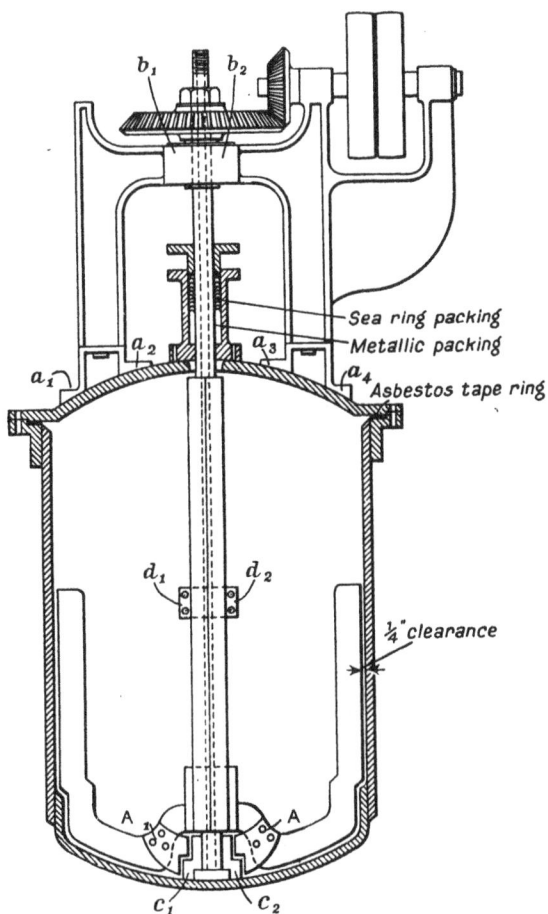

FIG. 15.—Sectional Elevation of Steel Autoclave.

from the cover flange to the beginning of the dish is from 5 feet 6 inches to 6 feet, while the depth of the dish in the centre is 1 foot. The height of the domed cover is about 10 inches and the internal diameter of the autoclave 4 feet 6 inches.

Such a pan has a capacity of about 700 gallons, which is a very suitable size for all-round manufacturing processes.

The flange on which the cover rests is either cast as part of the cylindrical body or riveted to it, although in autoclaves made of thicker steel it might be bolted into the body, the bolts being, of course, not tapped through.

For this type of pan it is sufficient to use steel of 1 inch thickness both for the sides and dish, although the cover would be thicker, say 2 inches. It is interesting to note that the joint made between the cover and the flange is of the " half-register " type, whereas up to now in previous models we have always dealt with full register joints. Fig. 16 shows the two kinds of joints on a larger scale, from which it will be seen that the full register would be selected if very high pressures were to be withstood. For this type of autoclave the half-register joint, if properly packed, is on the whole satisfactory. In this connection it is interesting to note a suggestion which appeared as a contribution of Mr. F. Bloor in the *Engineering World* of May 7th, 1921. In this it is shown that the pressure of the bolts on the half-register joint tends to squeeze the packing material out at *A* (Fig. 16 *a*) and causes a hinge movement at *B* to be exerted

by the pressure in the pan. The suggestion is that
the faces of the metal to be joined by half register
should be at an angle of 15° with the horizontal,
and that a clearance portion *C* (Fig. 16 *c*) should be

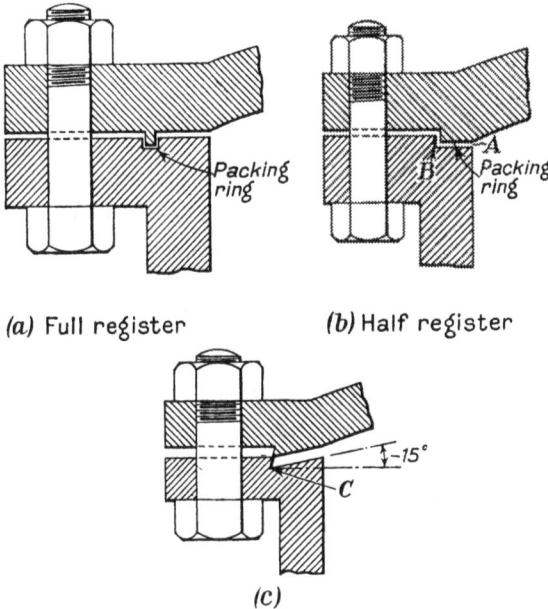

(a) Full register   (b) Half register

(c)

FIG. 16.—Examples of Joints.

provided. The effect of screwing down will now
cause the faces to find their true level, besides
which the ring can give to the point *C* and fill the
space allowed. Having made such a joint the effect
of pressure in the pan will be to cause hinge action
into the ring, thereby making the joint more secure.

The author has never personally tried such a device, but it appears well worthy of trial, especially in autoclaves working up to 30–40 atmospheres pressure.   To return to the description of the low pressure agitator autoclave, it must be noticed that the nuts and bolts which secure the half-register joint between the body and the cover of the pan are of massive construction and placed as closely together as is possible.   It has been stated that the internal diameter of the pan is 4 feet 6 inches, and so, allowing 1 inch thickness of steel and a reasonable width of flange, the total over-all diameter of the cover is about 5 feet 6 inches.   Hence, if these bolts are placed five or six inches apart that will mean between thirty and forty altogether, which can be taken as a reasonable number to secure a good joint. The steel superstructure, whose function it is to support the agitator shaft and driving wheels, cannot be cast in one piece with the cover, as is the case with smaller autoclaves, but is bolted on to the cover as shown in the figure at a_1, a_2, a_3, a_4.   The holes into which these bolts go must not be tapped through the cover, which is one reason why a liberal thickness of metal is allowed in the cover. In a similar manner the base of the stuffing-box is bolted on to the cover, the tapped holes again being only allowed to penetrate an inch into the metal. Attention is drawn to two points of great interest which can be seen both in Fig. 15 and the frontispiece.   These are, first, the height of the stuffing-box, which is considerable, and, secondly, the liberal space allowed between the top and sides of the

stuffing-box and the steel superstructure which
bears the agitator shaft. These are essential
features characteristic of good design in autoclaves,
for a large stuffing-box, properly packed, means a

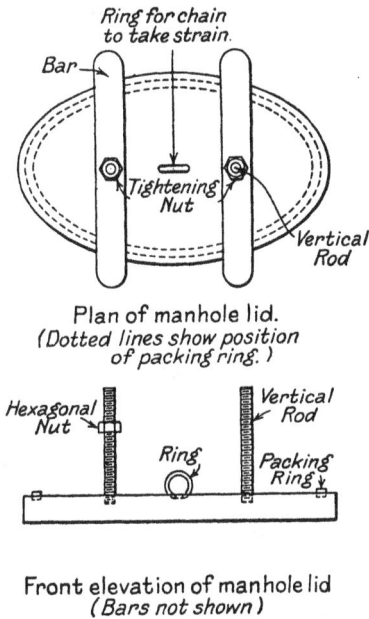

Plan of manhole lid.
*(Dotted lines show position
of packing ring.)*

Front elevation of manhole lid
(Bars not shown)

FIG. 17.—Cover of Autoclave Manhole Lid.

tight gland, while liberal space round it ensures
freedom of movement when it is necessary to screw
down the collar of the stuffing-box when the pan is
under pressure.

Although it has not been possible to show it in the
diagram, the cover of the autoclave should have a

number of openings in it, of which the following are necessary for ordinary work (Fig. 17) :

(1) A large oval opening, known as the manhole.

(2) A tapped hole for the thermometer pipe.

(3) A tapped hole for the pipe leading to the safety valve.

(4) A tapped hole for the pipe leading to the pressure gauge.

(5) A tapped hole for the pipe fitted with a valve for " blowing off."

(6) A tapped hole for the pipe leading to the compressed air supply.

These will now be considered in turn and commented on.

(1) In such large pans as these a manhole is an absolute necessity, as the lifting of the cover is an operation involving the use of pulley-blocks, and means putting the autoclave out of commission for the best part of a week. Once a satisfactory joint has been obtained between cover and body, it should only be broken when it begins to leak through wear or for the periodic overhauling of the plant, say once every six months or even a year. The function of the manhole is, therefore, the charging of the material for the batch and the usual means of access into the pan for examination. Naturally the size of the manhole should be as small as possible, for the larger the hole the more difficult it will be to make a tight joint when it is closed. It is usual in 700 gallon autoclaves to have

the manhole of oval shape and just large enough for an average man to squeeze through. The manhole is closed by a heavy lid of steel as thick as the rest of the cover. In the middle is a stout ring through which chains can pass which bear the weight of the lid while the workmen are adjusting the packing ring. The manhole lid is larger than the oval hole it is designed to cover, and has, therefore, to be passed edgewise through the hole into the pan. On its outer side at the extreme edge is placed a flat packing ring about an inch wide, made of black-leaded asbestos, which fits into a shallow groove designed for the purpose. Two stout rods cut with screw threads are fastened at right angles to the surface of the lid, and these thread through stout bars which bridge across the manhole itself. On screwing down two massive nuts, one on each rod, the manhole lid is pressed *upwards* firmly against the *inside* surface of the manhole, the asbestos packing ring causing a gas-tight joint to be made. It will be seen that, provided the joint is good, the effect of pressure will be to push up the manhole lid more firmly against the inside of the cover. This explanation is necessarily rather involved, but a study of the frontispiece, which shows clearly the nuts and bars holding the manhole lid in place, together with Fig. 17, which shows a sketch of the manhole lid from different positions, should help to make this explanation easier to follow.

(2) The thermometer pipe is not shown in Fig. 15, as it was desired to illustrate mainly features of

difference between works and laboratory pans. The general construction and function of the thermometer pipe are just the same in these large autoclaves as in the laboratory ones. It is a narrow tube, say of 1 inch bore, of thick-walled steel piping wide enough to conveniently accommodate the works thermometers, and is arranged to reach as near the bottom of the pan as possible. As there is generally plenty of room on the cover top, the works thermometer pipes pass vertically downwards, rather than at a slant, as was so often the case in laboratory apparatus. There is no particular virtue in this except that it is probably easier to avoid breakage of the thermometer if it comes out vertically from the pipe. A few words at this point on works thermometers suitable for autoclaves may not be out of place. The works thermometer differs from the usual laboratory one in being of great length—say from 3 feet 6 inches to 7 feet, the size of the bulb, bore of capillary and quantity of mercury inside being arranged so that the scale from 0° to 200° or 360° C. occupies the top 12 inches or 18 inches of the instrument.

The best works thermometers, especially for high temperatures, are filled with nitrogen to prevent the condensation of mercury vapour in the cool upper parts. Many of these instruments have engraved scales inside the tube rather than the graduations simply etched on the outside of the glass. If these inside scales are reliable and rigid this type is the better, as they are so much easier to read in the shadows round the pan. Moreover, they never alter, whereas the other type become attacked by fumes which destroy the black markings in the etched lines. In this connection it is useful to note that an outside scale thermometer whose markings have become so faint as to be almost invisible may be brought back to usefulness by rubbing with a cloth containing a little finely-powdered copper oxide, manganese dioxide or charcoal, which fills up the fine lines and makes these and the figures visible once more. In choosing a works thermometer for any particular autoclave attention must be paid to the length, so as to have the scale just nicely clear of the top of the thermometer pipe. Another even more important point is to see that the amount of immersion is reasonably near to that marked on the thermometer as corresponding to the graduations. With a large autoclave such as the one we are considering a 6 to 7 feet thermometer will be suitable, and these are generally graduated for 4 to 5 feet immersion, which will represent the amount that will be actually immersed in the hot oil when working with a full-sized batch. It is hardly necessary to note that the amount of oil in the thermometer pipe should be such as to fill it conveniently when the batch is on temperature—ordinary machine oil or high boiling paraffin form a suitable medium for ensuring that the thermometer is registering a true and accurate measure of the temperature of the batch.

(3 and 4) It was mentioned previously that the question of providing an adequate safety valve for autoclaves was not easy of solution. Never-

theless it is essential to have some such device, and
although the ordinary steam safety valves are not
very satisfactory for autoclave work, while a cap
at the end of a blow-off pipe designed to rupture at
a definite pressure limit is somewhat crude, there
are other neater and fairly effective devices in
technical use. As this is more a problem of the
high pressure works pan, it will be considered when
dealing with that piece of plant. Again, although
many autoclaves are merely fitted with straight
pipes of 1 inch bore leading direct to steel-tube
pressure gauges, there are one or two refinements
worthy of note, particularly relating to high pressure
large-scale work on which a few comments should
be made at the same time.

(5) When one speaks of the " blow-off " pipe or
valve in connection with works autoclaves, one
does not refer to the ejection of the contents of the
pan, which operation is more commonly described
as " blowing over the batch." The blow-off pipe
leads from the cover to the outside of the shed and
is, therefore, generally sloped upwards at an angle
of 60° with the horizontal, and passes out of the
shed through the wall or roof. It is controlled by
a good gland valve, for leakage here is just as bad as
anywhere else. The function of this blow-off pipe
is two-fold :

> (a) to give a means of releasing pressure in
> an emergency,
> (b) to blow off steam or other vapour and
> thereby greatly facilitate the cooling rate of

the contents of the pan at the end of the operation. Naturally this can only be employed if the product is not volatile in steam or the vapour concerned.

It was stated that this blow-off pipe was led out through the wall or roof, but if this wall is near a road or passage along which men are likely to pass the probability of condensed boiling water, etc., being dropped from the pipe must not be forgotten. In such cases, it is wiser to bring the blow-off pipe down again after leaving the shed and allow it to eject any liquids on to the ground or into a drain.

(6) The pipe supplying compressed-air service to the autoclave is in most cases a necessity, since it is usual to evacuate the pans at the end of the operation by " blowing over." The methods whereby the contents of the pan are blown by means of compressed air to the desired vat will be fully considered when dealing with the working of the plant. In any case, all autoclaves should have compressed-air service, as this forms a usual means of ventilation when the necessity for examination of the interior comes along.

The next items of constructional design to be considered are the agitator and agitator shaft. The latter can be either a rod or a tube of steel of approximately $3\frac{1}{2}$ inches or 4 inches diameter, the tubular shaft being the better for many reasons. Not only does a hollow shaft ensure less risk of bending under a heavy load, but it possesses advantages connected with the working of the autoclaves

4

which will be more fully elaborated in due course. The shaft is generally thicker inside the pan and narrows in diameter before entering the stuffing-box at the top end, and the setting in which it rotates at the bottom. After leaving the stuffing-box the shaft passes through the steel super-structure and is keyed or screwed on to the large bevel cog-wheel which forms part of the gearing arrangement by means of which the agitator is driven.

The diagrammatical illustration of the bevel cogs and fast-and-loose pulley for driving off the shafting is sufficiently clear to need no further comment save that it is usual, and desirable, to have all rotating parts working on ball bearings. A ball-race will therefore be provided in that part of the steel superstructure marked b_1 and b_2, Fig 15, in order to ensure smooth easy rotation of the large bevel cog-wheel.

As has been stated, the agitator shaft terminates at its lower end in a setting or cup (C_1 C_2—Fig. 15). This is bolted on to the dish, care being taken, as usual, that the holes are not tapped through any further than is necessary to give a really good, firm connection. It is essential that the bottom cup, the stuffing-box, and the hole through the upper part of the steel superstructure be absolutely in alignment so that the shaft fits tightly yet easily, and that no strains are set up when it revolves. Points like this determine whether an autoclave is going to be a faithful servant or a constant source of worry to the man who works it, as strains and

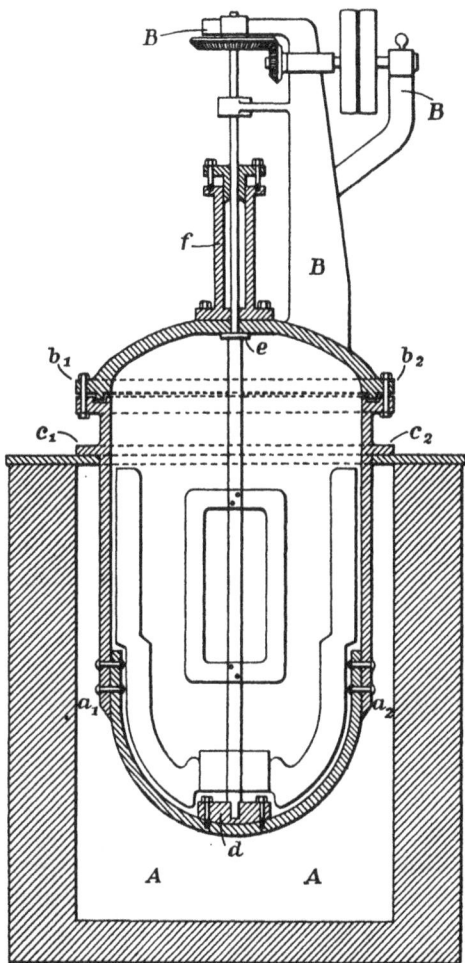

FIG. 18.—High Pressure Works Pan.

99

vibrations invariably lead to leakages under pressure.

Coming to the question of the shape of the agitator, there can be no hard-and-fast rule laid down, but the design shown is one that will be useful for most chemical reactions. It should be noted how perfectly and closely the agitator is made to fit the shape of the pan—$\frac{1}{4}$ inch clearance only being provided between the blade and the vertical side of the autoclave body. The anchor agitator shown is supplemented somewhat by the extra short blades d_1, d_2, Fig. 15, but a still more massive inside " gate " might with advantage be fitted for stiff mixtures, as shown in Fig. 18; the special shaped pieces of metal A, A (Fig. 15) are not usual, but were employed by the author for a special purpose, as will be explained later when dealing with the working of these pans.

As the heating of autoclaves is a problem which presents little difference whether the pan is of the low pressure works type (10 atmospheres) or high pressure works type (30 to 40 atmospheres), it is proposed to deal with the subject of the setting of autoclaves in brickwork or oil-bath as a chapter by itself.

A short description of the higher pressure autoclave will now follow, after which the problems of heating and working the pans will be considered.

THE HIGH PRESSURE WORKS
AUTOCLAVE

CHAPTER VI

THE HIGH PRESSURE WORKS AUTOCLAVE

WHEN considering this piece of plant it is recommended that the student refer back to one or two earlier chapters of the book. In the first place he should refresh his memory as to the meaning of high pressure when used in connection with works plant together with the probable size of such autoclaves.

Secondly, there are two types of autoclave either of which possesses some characteristics of the pan about to be described, namely, the high pressure semi-large-scale autoclave shown in Fig. 11 and the large low pressure works pan which has been dealt with so fully in the preceding chapter.

This recapitulation is necessary as, owing to the inevitable resemblance between the high pressure works autoclave and the two examples just cited, it would be very monotonous if an absolutely detailed description of this pan were set down, while, on the other hand, there are sufficient points of difference to make some comment on this type essential.

Briefly, then, we expect a full-sized works autoclave to have a capacity of at least 300 gallons, but as it has to withstand a higher pressure than the type shown in Fig. 15 it is probable that the capacity will be less than the 700 gallons quoted in the description of that pan.

Such is generally the case, and 400 gallons may be taken as a fair size for a high pressure works autoclave. As regards pressure, we have seen that,

although it is easy to obtain laboratory apparatus
to withstand many hundreds of atmospheres, even
with semi-large-scale plant of 20 gallons capacity
and upwards a working pressure of 500 lb. to the
square inch is considered high, and so a full-size
works autoclave of 400 gallons capacity and working pressure of 30 to 40 atmospheres may be
described justly as a high pressure works pan.

It is best to regard the semi-large-scale autoclave
of Fig. 11 and the low pressure works autoclave of
Fig. 15 as the prototypes of the high pressure works
pan, and so, in order to avoid useless repetition, only
a diagram and those points of outstanding interest
or difference will now be given.

Let us therefore study diagram Fig. 18, which is a
sectional elevation of a typical works high pressure
pan. In the first place, since the question of heating works pans, whether high or low pressure, will be
considered in a subsequent chapter, no method of
firing has been shown. The space A represents
where the coal or gas furnace for direct or oil-bath
heating would be situated. The general construction of the autoclave is very similar to that shown
in Fig. 15, but it is interesting to note that the
" dish " or bottom which is riveted on to the main
body at (a_1, a_2) is less shallow and may even be
hemispherical in shape in order to better resist
the higher pressures. Again, the overlap between
the " dish " and the autoclave sides is greater, a
double row of rivets being often employed. On
examining the top part of the diagram it is again
to be observed that there is a greater roundness of

design and less sharp curves or angles. These are small points, but they make for greater security and tighter joints under the higher pressures. The cover of the autoclave, which is more dome-shaped than that of Fig. 15, is fastened to the body by means of the nuts and bolts (b_1, b_2), the dotted lines showing the position of the great flanges of cover and body round which the bolts are placed as closely together as is consistent with convenient working. It will be remembered that a half-register joint [see Fig. 16 (b)] was perfectly satisfactory for the low pressure pan, but where pressures up to 40 atmospheres have to be resisted there is no doubt but that a full register joint made against a copper or lead ring is necessary. This is another point of difference between Figs. 15 and 18. The flange (C_1, C_2) shown in the diagram is not essential, but merely a method of supporting the autoclave and renders access to the bolts (b_1, b_2) easy.

It is seen that the agitator shaft is held in the pocket (d) as usual and passes up to the top of the pan, where it terminates in a safety collar similar to the design of that in the semi-large scale autoclave (Fig. 11). The tendency of the agitator shaft to rise when under pressure is naturally greater in these large pans when working at 40 atmospheres, and in this connection it is interesting to note that not only is the safety collar fitted at (e) (Fig. 18), but that the arrangement of the gearing is different from that of the low pressure pan. The effect of a slight rising in the shaft of the low pressure pan would be to grind the two bevel geared wheels

4*

together. Whereas in the high pressure pan, even if the safety collar were not fitted or worked loose, a lifting of the shaft would only react on the top of the superstructure (*B*) or would slightly loosen the contact of the bevel wheels.

The stuffing-box (*f*) is of similar design to that of the low pressure pan, save that it should be larger and could with advantage be water-cooled.

As regards the agitator, this is hardly a feature of the autoclave, but rather of the type of operation to be carried out in it : the one shown in Fig. 18 is a fine all-round system, being a combination of a close-fitting anchor to scrape the sides and a "gate" to mix the middle portions of a thick batch.

No pipe fittings are shown in the diagram, but the usual services described for the low pressure pan should be provided.

There is a tendency in recent models to substitute pyrometers for the thermometer and pipe, and if of reliable design this form of temperature measurement has naturally many advantages. For one thing, a number of pyrometers can be fitted which will show if the agitation is causing a truly uniform mixing of the contents, while, for another, automatic registration of temperature can be effected in the chemist's office during night shifts which enable him to know without possibility of error if batches have been kept to correct temperature for the specified length of time.

Still, with trustworthy foremen and good thermometers there is very little to be said against the older method of temperature registration by a thermometer in a thermometer pipe.

When dealing with the low pressure autoclave it was stated that, although the ordinary type of

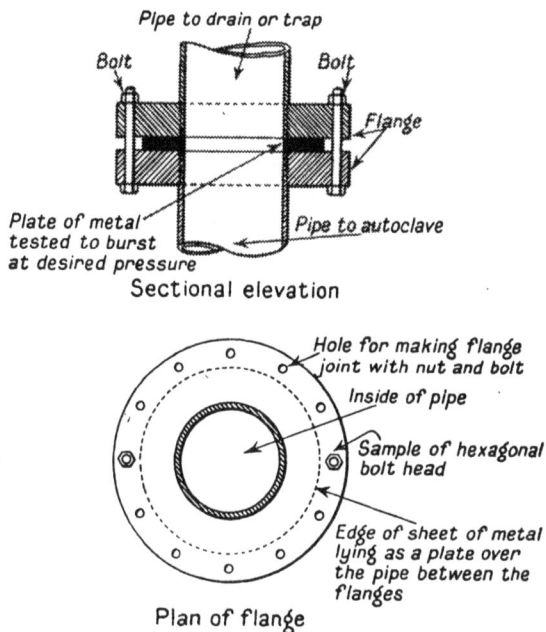

FIG. 19.—Safety Pressure Device.

safety valve was unsatisfactory owing to the likelihood of it sticking or getting made up with sublimed or distilled solid material, there were some fairly satisfactory safety devices to enable a blowoff of pressure to occur at a desired limit. Perhaps the most reliable is that shown in Fig. 19. This

consists in leading a pipe of fairly wide bore,
certainly not less than 1 inch, from the cover which
terminates in a flange joint made as usual with nuts
and bolts, the pipe continuing a short distance—
say to a drain—after the flange joint.   Between
the two flanges of the joint, besides the necessary
washer for withstanding pressure, there is a flat
plate of metal which is tested to burst at any
desired pressure.   The result of an excessive
pressure would be the rupturing of this plate and a
subsequent escape of the steam or gas.   Of course
some of the batch would probably froth up and be
lost, but that can hardly be avoided unless a
special trap chamber were provided into which
the pipe would lead instead of into the drain.   A
second point of greater interest with high than with
low pressure pans is the problem of the manometer,
or pressure gauge, as it is often called.   For
ordinary work there is no doubt that a good wide
tube looped in a complete circle and leading to a
steel tube pressure gauge is perfectly satisfactory,
but there is a device specially designed for those
cases where distillation or foaming causes blocking
of the leading pipe or the steel tube of the gauge
and consequent breakdown of the ordinary method
of pressure registration.   Fig. 20 shows this
arrangement in sectional elevation.   It needs very
little explanation except to point out that the effect
of pressure in the pan will be to force some of the
liquid mixture up the pipe (*a*), thereby compressing
the air in this pipe and forcing the oil of the oil seal
round the loop (*b*).   This will compress the air

again in that section of the tube nearest the gauge
(c), and so exert a pressure of clean air on the gauge
equal to the pressure in the pan itself. Unless
there are leaks the contents of the pan can never

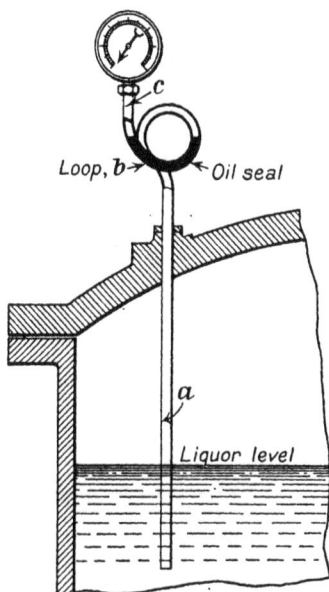

FIG. 20.—Special Pressure Registration Device.

get past the oil seal nor can the oil itself be driven
up into the gauge, as in each case there is this
elastic air cushion acting, so to speak, as a buffer.

Fig. 21 contains nothing new, it merely shows
how these last two devices can be arranged neatly

in order to use only one hole through the autoclave cover. The safety valve (*A*) can consist of the arrangement shown in Fig. 19 or any other type

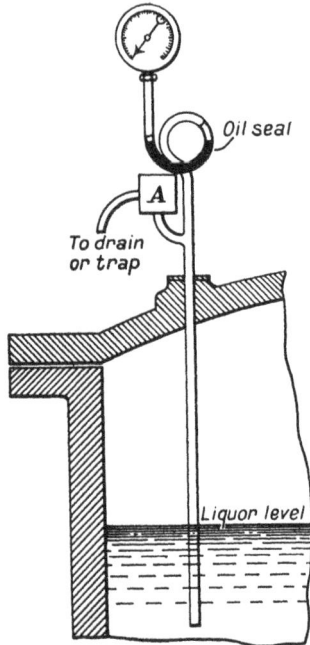

FIG. 21.—Safety Pressure Device and Special Manometer Connection.

considered to be suitable for particular require-ments. It should be noted that there is one objection to this particular combined fitting of manometer and safety valve, for in the event of the safety valve blowing, the contents of the autoclave will be driven by the pressure through the ruptured plate and down the pipe to the drain or trap, which will present the chemist with the unpleasant problem of getting a partially reacted and possibly solidified batch back again into the autoclave for the completion of the reaction !

THE HEATING OF WORKS AUTOCLAVES

CHAPTER VII

THE HEATING OF WORKS AUTOCLAVES

In discussing the heating of laboratory or semi-large-scale plant it was seen that there was very little to be said against oil-bath heating and much to be said in its favour. In certain cases, it is true, it was stated that oil-bath working was unnecessary and that gas firing, for example, with proper baffles to prevent burning was all that could be desired. Still, this was hardly a criticism of the oil-bath, but merely the expression of opinion that an alternate method of heating would meet a particular case. The problem of heating of works autoclaves is, however, a very different proposition, for there are factors operating here which do not come into play with other than real manufacturing conditions.

In the first place, the type of work done in works autoclaves is so different. Instead of a series of isolated batches carried out on a very small or moderately large scale mainly, if not solely, for experimental purposes, we are faced with the problem of routine manufacture in which it is desirable to get as much out of any plant in a given time as is possible. Then, again, the heating of laboratory or even semi-large-scale autoclaves of the smaller sizes by means of coal or coke fires is out of the question, as it would be impossible to regulate or control such tiny furnaces, whereas the coal or coke furnace is probably the most economical and general method of actually firing works pans.

115

If an oil-bath is employed it matters very little whether it is heated by fire or by gas burner, for, after all, it is the hot oil which really heats the autoclave. So to systematise this subject as much as possible we will consider the following three methods of heating in turn, pointing out the advantages or disadvantages of each and drawing conclusions therefrom :

(1) Heating by oil-bath.

(2) Heating by direct heat from coal or coke fire.

(3) Heating by direct heat from gas burner.

(1) As the name implies, this means that the autoclave is partially immersed in a bath of oil generally to within 6 inches of the cover flange so that the level of the oil is a little above the level of the charge. The type of oil chosen depends on the temperatures required, but a good thick black oil of coal tar or petroleum residuum origin is generally employed. The oil must possess certain characteristics. It must have as high a flash-point as possible—certainly well above the temperature to which it is required to be heated in order to give the desired inside temperature to the charge. In this connection it must be remembered that there may be a " lag " or difference in temperature between oil-bath and charge due to losses by radiation of as much as 40° C., so that if an inside temperature of 180° C. is required the oil should certainly not " flash " below 250° C. This presents, however, very little difficulty, since satisfactory

oils can be obtained which do not flash till well above the zone of most autoclave work. Another important property of an oil-bath oil is that it must be not easily "cracked," *i.e.*, decomposed when maintained at these high temperatures for long periods. It must resist oxidation and its viscosity and general physical properties must be unchanged after many rises and falls in temperature.

Of course, one cannot expect perfection and no oil can be used indefinitely, but as the emptying and refilling of an oil-bath is a messy and tedious business, this operation should not be necessary more than two or three times a year.

As regards the heating of the oil-bath there is little to require explanation. Whether heated by gas burners or fire, it should be done in such a way as to avoid burning the oil at the bottom and sides of the bath. After all, oil is an organic substance and must not be treated too roughly as regards heating. The chemist who never forces his baths, but nurses them carefully to the desired temperature, reaps the benefit in having much longer runs between dismantling and much less labour in cleaning up baths and autoclaves before restocking.

The one great virtue of oil-bath heating in which respect it is superior to all other forms is, of course, that the charge receives uniformity of heating and that the possibility of local burning is reduced to a miminum, if not entirely obviated. With the thickest mixtures, given reasonable agitation under this system of heating, the autoclave charge is at a uniform temperature throughout even during heating up and cooling down. A further advantage is that the use of an oil-bath will probably prolong the life of an autoclave, although whether the capital value thus gained compensates for the other costs associated with oil-bath work is a debatable question. Having stated these points, one has practically exhausted the list of advantages of oil-bath heating over the best forms of direct firing. There are, too, some disadvantages connected with the use of oil-baths for large-scale work which do not operate in laboratory or semi-large-scale.

The greatest of these drawbacks is the very considerable length of time inevitably taken in raising an autoclave to a desired temperature by means of an oil-bath, as compared with direct firing, and the corresponding length of time taken by the charge in cooling when the containing vessel is oil-jacketed. It may not always be possible to empty an autoclave immediately the reaction is over—perhaps the temperature and pressure are too high—which will necessitate the cooling down of the charge by, say, 50° C.

With a direct fired pan this would take, perhaps, three hours, but treble this time with an autoclave standing in a bath of heated oil. This disadvantage operates in two ways, and is more serious than appears at first sight. In the first place, it restricts the output of the plant so that a pan capable of making, say, five batches a week if direct fired, can only turn out three if oil-bath heated. In the second place, this greater period of time taken in heating up and cooling down is a serious departure from the conditions of any recipe which was in the first place worked out in a laboratory where the times of heating and cooling were very short. The use of oil-baths means often considerable modifications in recipe times and frequently means that recipes involving only short times can hardly be successfully imitated at all. As plant is nearly always used for routine production, the first objection is a serious one, which may mean that much greater capital must be sunk in plant for a given monthly output when oil-bath heating is used instead of direct firing.

Without wishing to lay down any hard and fast rule it is perhaps fair to state the case thus :

Where large output to the limit of a plant's capacity is not essential and where the substances involved are damaged if heated to temperatures outside rigid limits, and where the recipe is capable of much latitude as regards the length of heating, then oil-bath heating is clearly indicated. It is again desirable in works where plant depreciation must be reduced to its lowest possible state regardless of loss of output.

In most cases of routine manufacture, however, it is probable that the best forms of direct firing if carried out under experienced guidance possess advantages over the oil-bath method which exceed their drawbacks.

(2) It is not proposed to complicate the subject by going into details of the relative advantages of

coal or coke firing over gas firing in direct heated pans. Generally speaking, so much depends on the facilities available, but it is a fairly safe generalisation to say that of late years gas-heating appliances have improved so wonderfully that they are encroaching into fields at one time held entirely by King Coal. Still, most chemists or engineers faced with the problem of heating an autoclave will have to employ the coal or coke fire.

At first glance the problem might seem to consist of nothing more complicated than the building of a brickwork support for the autoclave and the provision of a fire-box beneath it. If that were done, however, there would be, it is to be feared, many charred batches in that autoclave and the log-book would show strangely variable yields from day to day !

Put as a general statement the problem of correctly and safely firing an autoclave by direct heat is solved by the provision of the following :

(1) A brickwork support for the autoclave, the top of which forms the staging which is on a level with the cover and manhole.

(2) A fire-box, well in front of the autoclave, with a firebrick arched roof leading to the space under and round the autoclave.

(3) Suitable " baffles " whose function is to block out too fierce heat rays and flames and to direct the hot gases into a definite path so that the whole of the autoclave is uniformly heated in a " hot air bath."

(4) Fire-box door and dampers so that the heating can be regulated to a wonderful degree of accuracy.

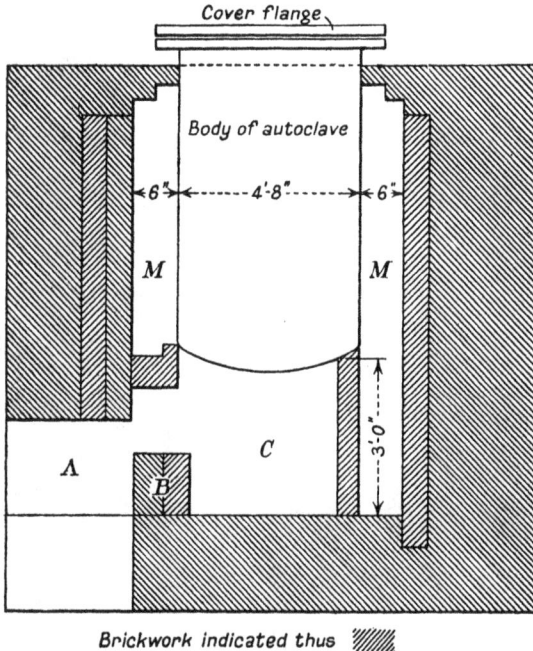

Fig. 22.—Arrangement for Direct Firing of an Autoclave.

A very general arrangement for direct firing of an autoclave is shown in Figs. 22, 23 and 24. It has been necessary to give these three sectional views in order to show all essential parts of the setting, and a study of these diagrams in detail is

recommended.   They represent, as has been stated,
a general arrangement and one that works fairly
satisfactorily for the firing of an autoclave containing a mobile mixture which is well agitated.
The method of circulating the hot gases and con-

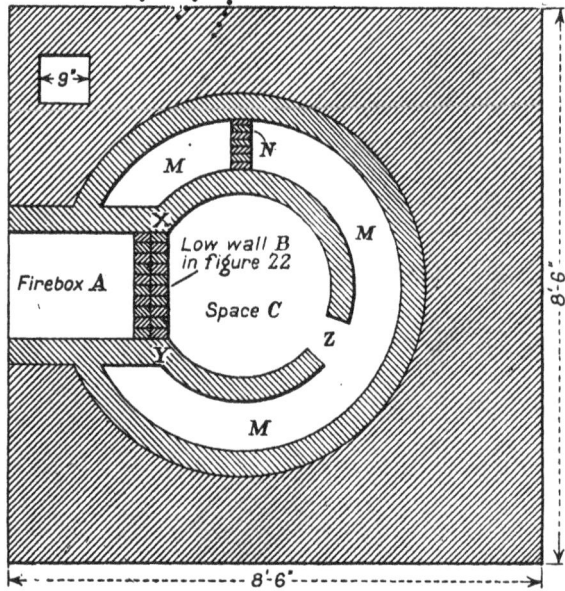

FIG. 23.—Arrangement for Direct Firing of an Autoclave.

sequent heating of the sides of the pan is good, and
although this setting possesses one grave defect,
this can be obviated by an addition rather than by
complete reconstruction so that the design deserves
description as it stands, especially as it is an
arrangement in very common use.   In order to

fully understand the design let us consider Figs.
22 and 23 together. The former is a sectional
side elevation and the latter a sectional plan as
seen from above.

The autoclave rests on a two-course brickwork

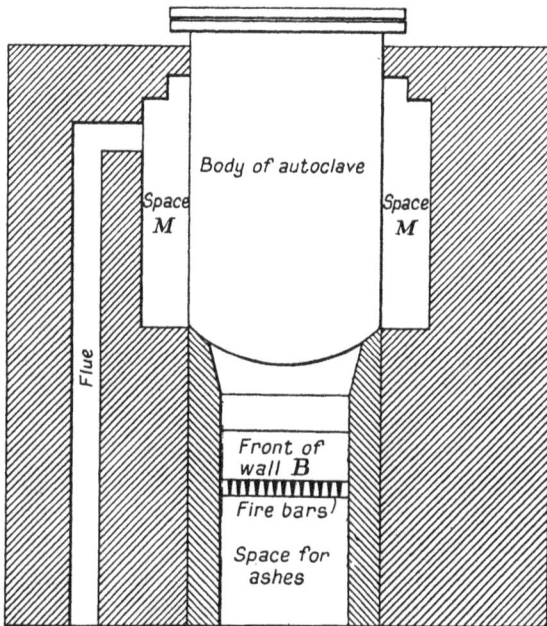

FIG. 24.—Arrangement for Direct Firing of an Autoclave.

wall which starts at one side of the fire-box tunnel
(*X*, Fig. 23) and passes round to the other side (*Y*)
as an unbroken circle save for a gap about 18 inches
across (*Z*). This wall is about 3 feet high. Some
further support could be given by allowing the

flange of the autoclave to rest in places on brickwork supports at the top, but that might make it
difficult to get at the cover bolts and is unnecessary
even with large pans. Turning to Fig. 22, we see
that the fire-box (*A*), which will be about 2 feet
in length, 18 inches across and a little more in
height, terminates in a short wall ('B) and leads
over this into the circular space beneath the
pan (*C*). The heat and flames from the fire pass
over the wall (*B*) into the space underneath the
pan, and from thence through the gap (*Z*, Fig. 23)
into the outer circular space (*M*) (Fig. 23). This
space is also marked (*M*) for the sake of clearness
in Fig. 22. The hot gases circulate all round the
sides of the pan, for the only outlet to the flue lies
over a baffle wall (*N*) (Fig. 23), which is also about
3 feet in height and serves to prevent too rapid
suction into the flue and consequent cooling of the
side of the pan near (*Y*). Having therefore had
free access under and round the pan the hot gasses
are allowed to escape into a flue 9 inches square
through a small hole and pass down this before
being finally led into a suitable chimney. Fig. 24
shows the front sectional elevation, and a study of
the three diagrams taken together with the above
explanation should make this rather intricate
matter clear.

Without being too specific it has been thought
advisable to indicate the relative sizes of various
parts of this setting by giving a few dimensions
both in the text and in the diagrams, and these
can be taken as roughly suitable for a large works

autoclave of seven or eight hundred gallons
capacity.

Such a setting could be used for heating oil-
baths if it was decided to place the autoclave in
one, although it would hardly be necessary to
baffle the gases quite so thoroughly in that case.

It has been stated that this arrangement pos-

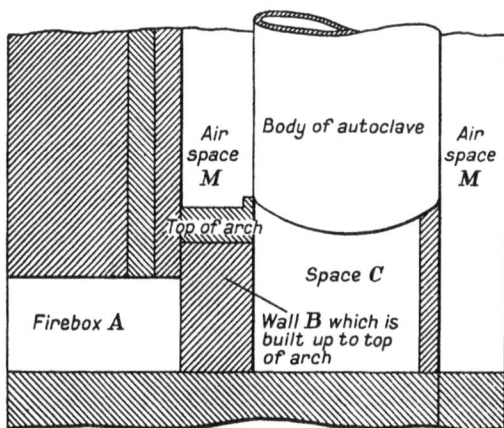

FIG. 25.—Showing " Baffle " between Fire-box and
Autoclave.

sesses one grave defect, and doubtless the discerning
reader will have detected it. Of course it is that
the flames will be drawn by the draught right into
the open space beneath the pan and will impinge
on the bottom, thereby causing burning of the
charge and deterioration of the plant. The author
has actually seen the bottom of an autoclave bright
red hot through too rapid firing in such a setting.

This defect, enormous as it seems, can be overcome by a very simple device. This consists in building up the wall (*B*) (Fig. 22) until it reaches the roof of the arch beyond the fire-box, not, of course, as a complete wall, which would cut off all heat from the pan, but as a baffle of " pigeon-holes." This is shown in Figs. 25 and 26, which are side sectional elevations and front elevation respectively. The

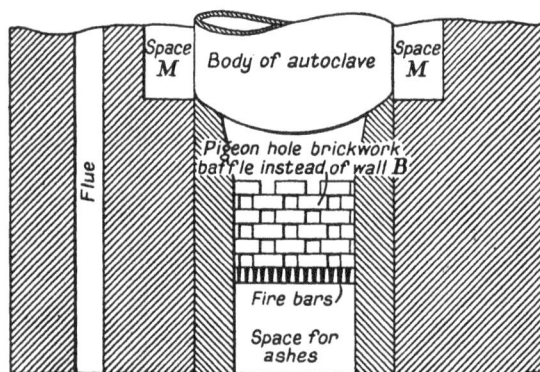

Space M Body of autoclave *Space M*

Flue

Pigeon hole brickwork baffle instead of wall B

Fire bars

Space for ashes

Fig. 26.—Showing " Baffle " between Fire-box and Autoclave.

effect of the erection of this baffle is remarkable, as with a good draught it does not damp down the fire, but entirely prevents the actual flames entering the space beneath the autoclave, which is to all intents and purposes heated by a hot air-bath. Such a modified setting, properly and skilfully worked, is an almost ideal method of firing pans intended for hard routine use, and one which allows the maximum output to be obtained from the plant·

Once the autoclave has been raised to the desired temperature it is surprising how small a fire will keep it steady, a skilled workman being able to regulate the temperature of the charge to within a degree or two by a combination of intelligent

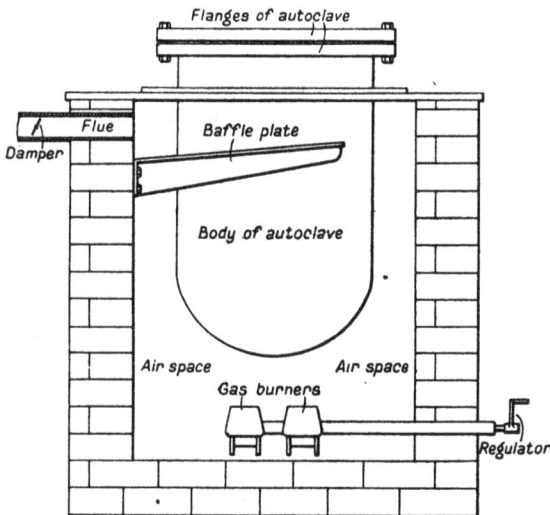

FIG. 27.—Sectional Elevation of a Setting to a Gas-heated Autoclave.

stoking and adjustment of the flue damper and furnace door.

(3) In conclusion, a few words must be said about gas-heated autoclaves. Even large pans of four or five hundred gallons capacity can now be heated by gas. It is usual to have the burners some little distance underneath the pan, heating

up being started by a small flame in order to avoid
burning. Gas heating can be regulated so finely
that the same elaborate system of baffles is not
required as is the case with coal fires, the simple
expedient of running a baffle plate three-quarters
of the circumference of the autoclave to force the
heated air to encircle the body of the pan being
sufficient. Although much of the regulation is
done by means of the gas supply taps, it is advisable
to have a damper in the flue as well, and this
enables one to keep the whole of the space round
the autoclave full of hot gases, which slowly circulate against the baffle and over it along the upper
part of the pan to the flue.

Fig. 27 shows a simple setting to a gas-heated
autoclave in sectional elevation.

THE WORKING OF LARGE-SCALE AUTOCLAVES

CHAPTER VIII

WHEN considering the use of laboratory autoclaves a few typical examples of reactions best carried out in such apparatus were given. It is not, therefore, proposed to go into the chemical side of working plant, as this must necessarily be similar to that already considered, but rather to pay attention to those details of plant usage in which large and even semi-large-scale work differs from the procedure of the laboratory.

A chemist who knows perfectly how to carry out a certain operation at his bench will meet with all manner of technical troubles when faced with the problem of manufacturing from that same specification on a large scale, so that it is the object of this chapter to anticipate to some extent these troubles and to show how they can be met and overcome.

The best method of so doing is to imagine ourselves in charge of an autoclave engaged in the manufacture of any of those important hydroxy- or amino-hydroxy-derivatives of naphthalene which are prepared by the fusion of a sulphonic acid with caustic soda. Such an operation will probably be carried out in the large low pressure agitator pan previously described, or, if the temperature of fusion were very high and the concentration of caustic soda low, whereby great pressures would be generated, in the rather smaller autoclave working up to 30 atmospheres. By far the most likely pressures would, however, be in the region of

7 or 8 atmospheres, so that the low pressure pan could be employed.

Before coming to the actual mechanical side of the subject a few words on the question of batch control tests would not be out of place. In order to get the best results from his plant the works chemist should keep a very complete record of all conditions in his log-book. The size of the batch should be the largest that the plant will safely take, remembering that no autoclave should be filled to more than 80% of its total capacity in order to allow for expansion of the liquid contents under heating.

In the laboratory the chemist has probably used a pure dry powdered sulphonic acid to melt, whereas in the works he may be delivered a paste containing not only water, but possibly also free mineral acid. In his calculation for the amount of caustic to give the required concentration he must, therefore, make allowance for the following :

(1) Caustic used in neutralising the free mineral acid (if any).

(2) Caustic used in neutralising any acid groups in the molecule of the starting material.

(3) The water formed by any such neutralisations.

(4) The water actually present in the paste of the starting material.

Having made any such allowances that the particular case may render necessary, he finally arrives at a figure for the caustic to be used and the extra water to be added.

Even if the process is a very steady regular one, a good control chemist will arrange to take samples at convenient stages for working up in the laboratory in order to get some approximation of yields. By this means if any stage of the process is not working properly—if the plant is leaking in any way—that defective stage will be shown up by the figures of the control tests and can be put right before any great financial loss has occurred. Whenever it is possible to keep individual batches separate this should be done, and the log-book should always show the yields of each batch throughout the week, month, or year. The author once had occasion to investigate a mysterious loss of yield in a process which up to that time had worked perfectly satisfactorily. Such a system of yield estimations by samples taken at each stage in the process showed the leakage to occur between a certain vat and filter-press, and it was found that a small hole had developed in the storage boiler sunk in the ground into which the suspension of the material was run before blowing it by compressed air through the filter-press. The effect of the compressed air was to blow a certain amount through this hole into the ground at each operation of filtration. This example shows how necessary routine tests are in order to discover defects before great loss has occurred.

To return to the manufacture of our hydroxy-derivative, let us suppose that the size of the batch has been fixed and all calculations have been made. The first problem is, therefore, to weigh out so much caustic soda and to melt it with the required amount of water. The caustic soda may be in the form of rock, delivered in thin sheet-iron drums, or in the form of powder or flake. If it is rock caustic then the drums must be split open by sledge hammer and the solid mass broken up to lumps not bigger than small loaves of bread. The powder or flake caustic can naturally be charged into the water just as it is.

Caustic soda is a dangerous substance to handle, and it is the duty of the chemist to see that his men are educated to this danger and protected adequately, even if, in their ignorance, they object to the precautions. In any operation involving the throwing about of caustic by shovel or the breaking of rock caustic by hammer, comfortably fitting goggles should be worn by the men concerned. Strange as it may seem at first thought, workmen often prefer to use rock caustic, although it involves the huge labour of breaking up, to the powder or flake form, the reason being that the shovelling of powdered caustic causes so much dust to get into the air, which leads to intolerable itching of the eyes, nose and skin. Perhaps the best form of all is a good dry flake which does not contain much dust.

If no liner is used in the autoclave it is essential that .the operation of melting the caustic with the requisite water be carried out in a separate pan, *not* in the autoclave. The fusion of lumps of rock caustic or even any form of caustic with a small quantity of water should never be carrid out in a steel vessel, as it attacks that material much more readily than it does cast iron. If the sulphonic acid paste contains much water, then the concentration of caustic in the extra water with which it is fused prior to charging the paste will be very high—perhaps 70%—and such a hot fusion will play havoc with a riveted steel autoclave. If this operation is, however, done in a cast-iron pan and the fusion blown over into the autoclave immediately prior to the charging of the paste, then very little harm is done. The blowing over of such a fusion does not sound a nice job, but if reasonable care is taken in making joints, and men are instructed to " stand clear " when the actual blowing is started, no harm need result.

In order to save time a small fire should be started in the furnace of the autoclave before the fused caustic is blown into it, as by so doing the brickwork is heated up and the pan itself made just reasonably warm. A little experience will soon show how much to heat the autoclave, the object being to prevent the sudden chilling causing the fused caustic to solidify on coming into the pan.

As soon as the caustic is in the autoclave the agitator is set in motion and charging of the sulphonic acid is commenced. This operation is carried out as rapidly as possible, care being taken to regulate the rate of charging and the firing up so as to have a steadily rising temperature throughout.

If the sulphonic acid is in the form of a paste
containing free mineral acid, or if it contains a
number of free sulphonic acid groups, then the
heat of neutralisation is considerable and helps
to compensate for the cooling effect of the mass of
material added.

Naturally it is impossible to fix details too
specifically, but for most operations of this nature
a large batch of three or four " pound molecules "
could be charged in from four to six hours, and the
temperature at the end should be from 120° to
130° C. Care must be taken in charging some
pastes to avoid adding in such sized lumps or so
rapidly as to prevent the rotation of the agitator.
This is known by the workmen as " scotching the
pan," and the problem of persuading the agitator
to start again in a badly " scotched " pan is not as
humorous as the name given to the disorder.
It sometimes means hours of patient leverage with
a long bar shaped to a blade like a cold chisel, with
which the bevel cogs are turned little by little.
When the agitator has thus been turned back, say,
half a revolution, the belt is slipped gently on to
the fast or driving pulley and by vigorous slapping
on of the belt one endeavours to persuade the
agitator to run round through the obstruction and
so start rotating smoothly once more.

Some rough treatment may be needed to cure a
bad " scotch," but too much vigour may lead to
stripped teeth on the gear wheels, or bent and
broken agitator blades.

There never was an operation to which the motto

" more haste, less speed," was better applicable than that of charging an autoclave with lumps of heavy hydraulically pressed cakes of sulphonic acids, but once again it is wonderful to observe the degree of skill which can be acquired by a really experienced workman.

As soon as the batch is charged no time should be lost in sealing up. This involves the fitting of the manhole lid, for it has been through the manhole, of course, that the charge has been added.

The manhole lid in large pans is a fairly massive article weighing a hundredweight or so, and the operation of fitting it requires just a little " knack." It is usual first to fit the wide flat blackleaded asbestos ring in its special groove on the top side of the lid, and then, by means of chains looped through the ring in the lid (see Fig. 17) and stayed over any convenient piping, to lower the lid until it hangs just over the manhole itself. One man will now take the strain on the chains, while his mate, by turning the heavy lid edgeways and sideways, will slip it through the hole. Once inside the pan it is adjusted to the correct position and then guided upwards until the asbestos ring is felt to be pressing evenly against the rim designed for the purpose. All this time the weight is being borne by the chains, but now one can slip one of the bars through the rods of the lid and spin the great hexagonal nut down to take the strain. The chains are removed, the second bar and nut fitted and the process of tightening up is proceeded with. The nuts are usually screwed up with box keys or

5*

spanners, tightening by means of a long lever bar being preferable to hammering up. The thermometer is now, *and only now,* placed in its oil pipe to avoid previous risk of breakage and all is ready to fire up to temperature. It is a good plan to educate the men to make a practice of clearing away all rubbish—old casks, papers, etc.—and to sweep the cover of the pan and the staging before firing up, as a neat, tidy staging helps to suggest that atmosphere of accuracy which is so necessary in autoclave work.

A chart should always be provided on which entries showing time, temperature (both of the charge and oil-bath if the latter is employed), pressure, and general observations should be made every half-hour. Apart from the men required for charging, one skilled hand can look after a number of autoclaves as they are coming up to temperature, especially if he has a stoker to fire up under his direction. This skilled hand will attend to the charts and watch ceaselessly for leaks developing in glands, manhole lid joints, cover joints or taps. He will also see that the oil cups on the bearings are working properly, for no one who knows how to get the best out of machinery will allow it to squeak or groan complainingly without investigating the cause and, if possible, supplying the remedy.

The time that it is necessary for the batch to be kept on temperature and, for that matter, the temperature itself, depends on the particular reaction which is being carried out. In the case of caustic fusions of naphthalene sulphonic acids we can, however, take from two to twelve hours as a reasonable range. At the end of the specified time it will be the object of the chemist to get the temperature lowered as quickly as possible. There are two main reasons for this. In the first place it is not generally advisable to keep the reacting bodies " on temperature " or near that temperature any longer than necessary, as the recipe has been worked out carefully to fix the time at which the best yields or purest products are obtained. Possibly longer time after a reasonable latitude will cause over-melting or the substitution of a second hydroxy-group for a sulphonic acid one, or, perhaps, it may cause darkening of the product. One should remember that a laboratory working is always carried out in a shorter time than a works one of the same recipe, so in order to keep as faithfully as he can to laboratory recipe conditions a works chemist must avoid unnecessary loss of time. The second reason for speed in lowering the temperature of the batch is an economic one. As will be seen, it is usual to empty the autoclave by blowing over the contents through a pipe fitted in a light manhole lid which is substituted for the heavy one. This means that the pan must be opened, and in order to do so all pressure must be released. Even in those pans fitted with permanent blow-over pipes it is not desirable to blow over strong caustic fusions at very high temperatures and pressures, so that some cooling and release of pressure will be necessary in any case before the pan can be evacuated. That was why this reason was described as an economic one, for an autoclave represents capital, and the longer it is occupied on one batch the less will be the monthly output from that capital.

The usual procedure with caustic fusions and many other autoclave operations is to draw the fire on completion of the reaction time and to open fully the fire-box door and flue, while the blow-off valve is cautiously released too. Naturally if the product is volatile in steam this latter course cannot be adopted, but wherever possible it should be done, for the rapid boiling off of some of the water has a marked effect on the cooling rate of the batch.

The degree to which this treatment can be pushed and the temperature to which the batch is cooled are factors which only experience in any particular recipe can fix, but as a general working rule one can say that caustic fusions can be safely and conveniently blown over at 120° C., and that any pan should be opened as soon as there is no residual pressure. The blow-off pipe does away with any trouble caused by residual pressure due to gases formed by decomposition, but to avoid the possibility of accidents the following procedure should be adopted in opening a large works autoclave. As soon as the blow-off pipe ceases to show emission of gas or steam and the needle of the gauge is at zero (which is best seen by striking the gauge a smart blow with the palm of the hand, when the needle should " click " against the pin near the zero mark), then the hexagonal nuts on the rods of the manhole lid should be loosened with a long-handled spanner—say a No. 6 or 7 adjustable. The heavy lid will probably remain stuck against the asbestos ring, but a smart tap with a long bar will cause it to ease, and once steam is seen escaping the chains can be fixed and the lid removed by exactly the reverse process to that described in fitting it.

A very common method of loosening the manhole lid, and one which, although it sounds dangerous, is probably perfectly safe in practice, is as follows. As soon as the charge hand sees that only a few pounds per square inch pressure is shown on the gauge, and while the blowing off still continues, he loosens the nuts of the manhole lid a considerable amount. This has no effect, for the remaining internal pressure holds up the lid against the manhole. The charge hand busies himself with other jobs, knowing that as soon as the pressure is released a characteristic crash will tell him that the lid has " dropped," *i.e.*, has slid down the rods just as far as he loosened the nuts. He then can attend to the removal of the lid and subsequent operations.

Once the heavy lid has been removed a light " blowing-over " lid is fitted. It is exactly similar in design and general construction to the heavy one, only much lighter, as it has only a few pounds per square inch of air pressure to withstand. In the centre of this lid is a round hole for the blow-over pipe, which fits down to a flange on the pipe, an air-tight joint being made with small nuts and bolts against a flange ring of rubber, composition, or asbestos. The blow-over pipe is shaped to go to as near the bottom of the pan as possible, and terminates at the top in a union joint which is made on to the permanent pipe lines in the shed. In a well-designed shed the pipe lines will be such as to give the chemist a considerable variety of vats into which he can blow his batches, and also some form of emergency receptacle into which he can put a batch as a last resort in case of complete breakdown of vats or pipe lines. When one remembers that the value of a batch may be several hundred pounds, a little thought and expense in planning for emergencies are not out of place. As an example of emergency receptacles one may quote boilers sunk in concrete pits and floored over. These are not expensive, are out of the way, and come in very useful as storage tanks for liquors and as temporary resting-places for batches waiting their turn if the plant is working at full capacity and the inevitable breakdowns occur.

To return to the blowing over of our autoclave contents. Having fitted the light lid, made good the blow-over pipe, and connected up to the permanent piping, then there is a golden rule that every chemist or foreman, no matter how experienced, should invariably follow. It is this : *Make sure that only those valves are open which lead to the desired destination of your batch.*

In a complicated tangle of piping it is surprising how easy it is to omit to close one of many paths.

The author has known of very many cases—some humorous, some tragic—where batches have been blown over into the most unlikely places, much to the surprise of the "blower" and "receiver," through the neglect of attention to that golden rule!

Again, then, let it be repeated—follow your pipe line from the autoclave to the vat, or to wherever be the destination of the batch, and see that all valves are closed save those required to direct the route. Then all taps on the pan should be watched, and finally, when everything is *known* to be right, give a warning shout to the men at the "receiving" end and turn on the compressed air.

There is some romance even in the life of a works chemist if one cares to look for it, and perhaps of all moments the one that gives the greatest feeling of pride is when one sees one's "first batch" passing smoothly and without hitch in a great stream full bore through 3 inch pipes, splashing triumphantly from the autoclave into the vat on the staging above.

It has been shown that considerable time must elapse between the completion of the reaction and the blowing over caused by the slow rate of cooling of the autoclave contents. This is particularly so in those cases where it is impossible to blow off steam by reason of the volatility of the product in that vapour, and most particularly so in oil-bath heated autoclaves used for such operations. In cases like these one may have to allow as much as twelve hours before blowing over can be carried out. Naturally such a disadvantage calls for investigation in order to see whether a means of saving this valuable time can be devised.

The greatest loss of time is caused by having to wait until all pressure has disappeared in order to change the manhole lids and fit the blow-over pipe, and it is in order to do away with this part of the process that autoclaves have been designed with blow-over pipes fitted permanently. Although in some cases they may work satisfactorily, yet on the whole this design has not been favourably received by most chemists. One great objection is that these fixed blow-over pipes are very apt to get silted up with solid matter and so to refuse to work at the desired time, while another drawback is the additional likelihood of leakage during heating.

Perhaps the most reliable and neatest way of reducing the time required between completion of the reaction and blowing over to a minimum is that of utilising the hollow shaft carrying the agitator blades as the blow-over pipe, and, provided certain precautions are adopted, this is a method that works satisfactorily, say, with nine batches out of ten, even in those cases where much solid insoluble matter is carried in suspension.

In order to explain this most accurately, let us consider two cases :

(a) The blowing over of a β-naphthylamine batch.

(b) The blowing over of a fairly thick caustic fusion, containing salt and insoluble matter in suspension.

(a) The points of interest here are that the temperature at the end of the reaction will be, say, within the region of 140° to 150° C. and the pressure about 7 atmospheres. If the mass is cooled down, when a temperature well above 100° C. is reached and before the pressure has gone, the β-naphthylamine will separate out in the form of lumps of solidified oil or small balls like marbles. Even if the ordinary blow-over pipe is fitted it is impossible to do anything until the batch is reheated up to about 120° C. to ensure complete melting of the naphthylamine.

Suppose, however, we have a hollow shaft with holes cut in the bottom to allow entrance of material and fitted with a good " gland " tap at the top above the gearing wheels, then if any suitable piping arrangements be made so that on stopping the agitator connection can be made by means of a union joint, it would be possible to blow over at the end of the reaction through the hollow shaft and gland tap into the pipe lines to the desired vat. This could be done either immediately, using the 7 atmospheres residual pressure as the blowing force, or, if this seemed too violent or 150° C. too hot, then one could wait a few hours until the temperature was, say, 130° C. and the pressure 4 atmospheres.

There are one or two precautions it is wise to take with β-naphthylamine or any similar preparation. If one starts blowing through cold pipes, the first lot may solidify in the gland tap, or even in the pipe lines themselves on a winter's day, especially if there is some distance to travel from autoclave to vat. It is necessary, then, to have steam supplied so that the whole pipe line can be steamed out immediately before opening the gland tap and letting loose the batch. Again, suppose the gland tap has leaked *very slightly* during the twelve, or twenty-four, or even thirty-six hours required for amidations. The effect will be a gradual rising of the molten contents of the pan until they reach the cool portions of the shaft, wherein they will soldify, thus rendering it impossible to blow-over this way. There are two things which the " old hand " can do to guard against this besides the obvious one of overhauling the gland tap before each batch. The first is to close this tap *before* the pan is charged, thus imprisoning air in the entire length of pipe from the gland tap to the bottom of the hollow shaft. Under pressure this elastic cushion of air will only allow the batch to rise slightly in the hollow shaft, and there is a chance that even with a small leak on the tap the batch will not have risen up at the end to the cool parts of the shaft. The second precaution is to have a good blow-lamp handy, and to play on the top part of the shaft and the tap itself with the flame just before blowing over, so as to melt any of the product that may be suspected of having been forced up through leakage. This sounds rather drastic treatment, but, if done sensibly so as to gradually warm up the offending part to a temperature above the melting point of the product, it is surprising how successful a trick of this

description can be. In the case of β-naphthylamine, if the temperature is about 145° C. and
pressure 100 lb. per square inch, and if the procedure
just described be carried out and ice and water
placed in the receiving vat, it is perfectly safe to
connect up and blow over immediately the proper
amidation time is completed.

It is curious how prejudiced many chemists are
against blowing over in this way with a residual
pressure, but there is no risk if reasonable care be
taken, and an immense amount of time and worry
is saved.

(b) In the case of the caustic fusion there is not
the problem of the solidifying oil to consider, but
we are probably left to deal with a batch at a very
much higher temperature on completion of the
fusion. This will have to be blown to a vat and
made acid, and in this connection it is interesting
to note that a considerable quantity of the acid
necessary for neutralisation can safely be placed
with ice in the vat, provided the batch be blown
over slowly and steadily. It would hardly be
advisable to blow over a caustic fusion at, say,
200° C., so with the hollow shaft arrangement
the procedure on completion of the fusion time
would be to draw the fire, open the flue, and start
blowing off steam through the blow-off pipe. The
agitator must be kept going right up to the moment
of connecting up the hollow shaft to the pipe lines,
which operation is done as soon as the temperature
has sunk to a more reasonable figure, say 140° C.
The pressure may not be quite enough to send the

batch over—everything depends on the concentration of the caustic and the height to which the batch is to be lifted—but it can be supplemented with compressed air if required. The precaution of steaming through the pipe lines should never be omitted. In some cases where much solid is in suspension it is found that the holes at the bottom of the hollow shaft become " made up." This is a troublesome fault, but can largely be obviated if care is taken to ensure that the space round these holes is well agitated. On referring back to Fig. 15 attention is called to the plates A, A_1, the special function of which is to keep the inlet holes of the hollow shaft clear. With these extra fittings and the careful following of the hints given when considering the blowing over of amidations there is no reason why hollow shaft blowing should not be much more common than is the case.

ROUTINE RUNNING OF LARGE-SCALE PLANT

CHAPTER IX

IN the preceding chapter the actual working of a large-scale autoclave was elaborated in some detail with a view to provide hints which might be helpful to the inexperienced. Thus, a typical autoclave manufacturing process was taken and considered step by step from the charging of the starting material to the blowing over of the batch. It was not considered desirable to go into the details of the subsequent working up of the batch, for these come under the heading of general chemical works processes rather than high pressure work. It is proposed now to go into the question of the working of autoclaves from a rather broader viewpoint than heretofore. By this one means the management of high pressure plant rather than the mere running of one isolated autoclave for one particular batch, and although this is naturally a subject on which it is difficult to generalise, there are, nevertheless, certain rules the elaboration of which may prove helpful.

Let us first consider the case taken in the last chapter—namely, that of a chemist in charge of one autoclave who is required to turn out a maximum output of a hydroxy-derivative obtained by the fusion with caustic soda of a sulphonic acid. Such substances are very common in chemical industry, particularly the intermediates for dye-stuffs of the naphthalene series. Output from plant is almost as important as the " yield " of a process. The former represents so much more

151

money coming from capital invested, while the latter is, of course, the term used to indicate the production from a batch calculated as a percentage of that theoretically possible. It is not too much to say that more skill is required in a works chemist to squeeze the greatest possible output per week from his plant without injury to it than to obtain good yields from the process he is working.

In the case of the chemist with one autoclave, much will depend on whether he has reasonable shift facilities for working the plant all through the twenty-four hours of each day, but it is generally desirable to make the following arrangements.

There will be a large shift working during the hours that the chemist is himself present at the factory and during which all the essential parts of the process should be done. The remainder of the time, say sixteen hours, will be divided into two shifts, each of which has the services of just as many men as are needed to keep the process going. With experienced men, a charge hand and two good labourers should be able to attend to three autoclaves during the day shift, while two men on each of the night shifts will be sufficient. The object which should be aimed at is one batch per twenty-four hours from the pan—Saturday morning coming in very useful for cleaning up, boiling out the autoclave, and a general winding up of the week's work. This will, of course, be possible only if the process requires a short time " on temperature," for it is obvious that if the batch has to be maintained at a certain temperature and pressure for, say, twelve hours, the whole of the other operations could not be squeezed into the remaining twelve hours of the day. Again, with oil-bath heating, the ideal " one batch per pan per day " is hardly practicable, but with good management, good workmen and a short-time fusion it is possible if a direct-fired pan is used.

Let us, therefore, consider a typical time-table. On arriving at the works at, say, 9 a.m., the chemist will find the previous day's batch in the pan, the fire drawn, and the pressure blowing off. The morning will be spent in getting ready the material for the day's batch, breaking caustic, and weighing out quantities of the sulphonic acid, which will probably be in the form of moist cakes or dried powder.

All this material will be got on to the stage, and if it is the custom to melt the caustic in a separate pan, this can be charged and the melting started. The men concerned with the working up of the previous day's batch will have all in readiness for the time when it is to be blown over. This will probably be about noon, the operation taking place preferably through the hollow shaft by means of 40 or 50 lb. per square inch residual pressure. As soon as the old batch is out of the pan, the water and caustic for the new one are charged, or the melted caustic is blown in, as the case may be, and a fire started under the pan. Charging of the sulphonic acid is now begun, which operation may have to be done carefully to avoid " scotching " of the agitator, and this will certainly take the remainder of the day. The fire is so maintained that the temperature is rising all the time, ending at 125° C. by from 4 to 7 p.m. The manhole-lid is now fixed, and the heating up to the desired temperature started. This must not be done *too* quickly; but, without forcing the firing, the batch should be on temperature between midnight and 3 a.m. If the time of heating on temperature is up to four hours we may expect the fire to be drawn not later than 6 or 7 a.m., which brings us round the cycle to another day's routine.

This means very hard work all round and allows no latitude for the inevitable breakdowns, but it serves to show what can be got out of a pan if maximum production is essential. A much more comfortable arrangement would be to have two pans working on alternate days, in which case the programme would be as follows :—

Starting with two empty pans.

Monday.—Charge caustic in No. 1 pan and get this melted as quickly as possible. Charge material in the afternoon and have the pan sealed by 5 p.m. Heat up and get " on temperature " by Monday night. Also charge caustic in No. 2 pan on Monday night and have this slowly melted during the night.

Tuesday.—Blow over batch from No. 1 pan when ready, and re-charge this pan with caustic for slow melting on Tuesday night. Make batch in No. 2 pan during the day, as the caustic would be ready melted the first thing in the morning.

Wednesday.—Blow over the batch from No. 2 pan and recharge with caustic for melting on Wednesday night. Make batch in No. 1 pan.

Thursday.—The same as Tuesday.

Friday.—The same as Wednesday.

This plan, of course, only means one batch per alternate day per pan, but it has the advantage of giving much more time for the processes, while it is quite possible to keep to this programme even if the time of heating on temperature is as long as twelve hours. Since we started with empty pans, Monday was rather a rushed day, but even if No. 1 pan was not sealed until late at night it would not matter, as that pan was not required for actual manufacture again till Wednesday.

This should serve to show how autoclaves could be worked, but, of course, each chemist must be prepared to plan his programme to suit his own special needs. Maybe in some cases a modification of the above schemes could be employed, in which three batches were made in two pans in two days, but this would not be so good, as the times for definite operations would change from day to day. The best programmes, and those which suit the workmen most, are those in which each man can have his own definite job to carry out each day at roughly the same time.

When dealing with the routine running of larger numbers of autoclaves, say six or ten, so much again depends on the processes that it is difficult to lay down any general principles likely to be of help. At the same time there are certain rules which have received the sanction of experience and which it will be, therefore, advisable for a chemist to follow in most cases. In the first place, it is often a good plan to group the pans in pairs and so arrange the work that batches are made in Nos. 1, 3, 5, pans on one day and in Nos. 2, 4, and 6 on the day following. Again, although the pipe fitting of the shed will be such as to fall in with this system of groups, it should, *even at the risk of becoming intricate,* be such that the chemist in charge can blow, not only from *any* autoclave to *any* vat, but also from one autoclave to another. The works chemist should always remember that breakdowns seldom occur at convenient times, but, generally speaking, machinery most often gives under full load. It is only natural that it should be so, and the really good works chemist is the one who has provided, not only for smooth running, but also for every conceivable disaster. The author remembers a case in which a bad leak developed in the riveting of the dish of a large autoclave to the body just as the batch was coming up to temperature and pressure. The mixture was being forced out of the pan and dropping into the furnace. It was a large and valuable batch, but in less than ten minutes a few union joints had been made, a few taps opened and shut, and the batch was blown by means of its own pressure through the hollow agitator shaft into an adjacent autoclave, the total loss being less than 2%. This was an instance of foresight and good management.

This brings us to another point—whenever working with a large number of pans it is always wise to " carry a spare." This does not mean that one pan must remain idle for most of the time on the off chance of an accident occurring, but simply that the whole plant is not being worked to the absolute limit of its productive capacity. All sorts of set-backs are liable to occur, and even the best of time-tables cannot be rigidly followed. If there is some latitude allowed there will be a balance about the whole installation which will be reflected in the regularity of the monthly production returns. Perhaps a batch may be started in No. 2 pan, but owing to a blockage in the pipes it cannot be cleared ready for the correct time. If there is no spare autoclave, then either a batch must be " dropped," *i.e.*, not made on that day, or it will be started late, which will mean disorganisation and hurry, bringing as its inevitable result a mysterious drop in yield. There is no doubt that the regularity of production and yield from an installation which " carries a spare " more than compensates for the capital which on the face of it is standing idle for some part of the week. While dealing with the question of spare parts of an autoclave plant, it should be noted that the remarks made are particularly applicable to the wooden vats into which the charge from the pan is so often blown. The life of these vats is never very long where corrosive liquids are employed, and as they are comparatively inexpensive a reasonable excess over actual requirements should be provided if regularity of production is desired. Here again the chemist should remember that a vat will often develop a bad leak when empty, which will become apparent just as his batch has blown over, and, therefore, arrangements should be made for passing the contents from one vat to another. This can be done either by some form of ejector worked by steam or by having an arrangement whereby any vat can be emptied into a storage boiler and the batch blown up again into any other vat.

The output of a large number of autoclaves depends on a great number of factors, not the least of which is the general design of the shed. Much time can be saved if the staging built round the autoclaves is of liberal size to accommodate the materials for the batches, and is adequately served with hoists to enable both men and material to reach the pans by the most direct routes. Again, there should be plenty of space round the pans, so that all valves—safety, blow-off and blow-over— are easy of access. A man crouching in a corner at the back of an autoclave on a narrow stage against a wall would be caught in a veritable death-trap if the ring of the manhole-lid blew out under pressure.

In conclusion, mention must be made of the routine testing of autoclaves in order to ensure the safety of the men employed in working them. This is a very important part of a works chemist's duties, for neglect of his pans not only causes rapid deterioration, but thereby incurs the possibility of explosion, which might entail the destruction of an entire shed and loss of life.

As has been stated previously, a wise chemist will always take any opportunity that arises of boiling out his pans with water or weak soda ash solution in order to prevent the formation of crusts of burnt matter on the sides and bottom.

Some regular plan should be adopted for the cleaning out and examination of autoclaves, and a record of the result of these examinations entered in the log book. This examination of an informal nature can be undertaken by the men working the plant; in fact, a charge hand who is keen on his job and fond of his pans will welcome it. The moment anything out of the ordinary—an indentation or signs of splitting in the riveting of the dish to the body of the pan—is observed, then the plant should be handed over to the engineers for skilled investigation.

Whenever any man is about to enter a pan through the manhole there are certain rules which should be carefully followed :—

(1) The pan must have been boiled out until absolutely free from dangerous chemicals.

(2) The pan must be reasonably cool.

(3) The compressed air must have been blown through the pan for some time, and it must be proved that the air inside will support combustion, *i.e.*, of a candle flame.

(4) *The belt operating the stirring gear must have been cut off.*

160 *Autoclaves and High Pressure Work*

This last is of extreme importance. It is not enough to slip the belt on to the loose pulley, or to stop the motor driving the shafting. Someone may come along ignorant of there being a man inside the pan and thoughtlessly start the agitator. It is far better to insist on the lacing of the belt being cut before a man is allowed inside an agitator pan. Another good plan is to insist on men engaged on working inside autoclaves, whether for examination or actual repairing, always being in pairs. After all, no matter how one tries to make it safe and easy, working in an autoclave is bound to be a cramped, hot, and trying job. Even a strong man may suddenly become faint, so that as a reasonable precaution he should have a mate standing by in case of need.

Every few months, whether there appears to be anything wrong with the pan or not, it should be tested and examined by the maintenance engineers. The testing for pressure will consist of subjecting the pan just as it stands, with all valves, gauge, and agitator, to hydraulic pressure. This is done by pumping water into the pan through any suitable opening—say a valve provided for the purpose on the pipe to the blow-off or pressure gauge— until a definite desired pressure is indicated on the gauge. Often the pump used will have its own gauge, so that this constitutes a test for the pressure gauge as well as for the autoclave itself. This operation is perfectly safe, for if the manhole-lid ring or any other part gives before the desired pressure is reached it only means at worst a shower-bath for the operator and some amusement for the onlookers.

As regards the pressure to which pans should be subjected in this test, there is some difference of opinion. The author firmly believes that an autoclave should be hydraulically tested up to the full figure given by its makers, *but only worked to half that pressure.* Thus if a pan is supposed to stand 300 lb. to the square inch it should always be tested up to that limit by this hydraulic method, but no process evolving a gas pressure above 150 lb. to the square inch should be worked in it. After all, hydraulic testing cold is not nearly so searching as the pressure from steam or ammonia or ethyl chloride vapour at a high temperature, and the system of working at half the tested pressure allows a reasonable margin of safety.

In addition to the testing of autoclaves instituted by the works chemist, most pans are periodically examined by boiler inspectors or, if insured, by the officials of the company with whom the policy has been taken out.

6

INDEX

INDEX

166   *Index*

www.ingramcontent.com/pod-product-compliance
Lightning Source LLC
Chambersburg PA
CBHW031359180326
41458CB00043B/6546/J